In the Garden

Essays on Nature and Growing

In the Garden

Essays on Nature and Growing

DAUNT BOOKS

First published in the United Kingdom in 2021 by
Daunt Books
83 Marylebone High Street
London W1U 4QW

2

A CIP catalogue record for this title is available from the British Library.

ISBN 978-1-911547-92-1

Typeset by Marsha Swan
Printed and bound by TJ Books Ltd, Padstow, Cornwall

www.dauntbookspublishing.co.uk

Contents

The Garden Remembered

The Collective Garden

The Language of the Garden

The Sustainable Garden

The Garden Remembered

The Gardening Eye

PENELOPE LIVELY

The urge – the compulsion – to garden is genetic, so far as I am concerned, and runs down the female line. My grandmother was a skilled gardener and created a large garden in West Somerset, its landscaping and content much influenced by William Robinson and Gertrude Jekyll. My mother, her daughter, made an English garden in Egypt, complete with topiary, lily pond with weeping willow, a pergola walk. I have gardened more humbly but with equal enthusiasm, and my own daughter made herself long ago more knowledgeable than any of us by acquiring Royal Horticultural Society qualifications. And now one of her daughters is showing significant interest in her window box.

I grew up in that garden in Egypt – literally, because much of life was lived out of doors in the North African climate. I communed with a eucalyptus tree, sat reading in a lantana bush hideaway, swung from the aerial roots of the banyan. The structure of that garden, its sights and sounds, are sunk deep into my psyche and have a lot to do with my own life in the garden, I am sure. This began way back when I was first married, and we found ourselves the proud possessors of a small suburban back garden. Neither my husband nor I had ever laid hands on spade or trowel before; we were in blissful ignorance of what to do, but we set to and cleared out a bed in which to plant the little bright green rosettes of some plant we saw coming up all over the place. My grandmother visited, and eyed our creation with amazement: 'Why on earth have you planted out all that willowherb?'

Well, that's how you learn, and learn we did, over time. Our gardens got progressively larger as we moved from one place to another, and ended up with an Oxfordshire garden that had two streams running through it and a vegetable area that had been a farmyard for four hundred years – the soil was so rich that we grew vegetables in industrial quantities. But for both my husband and myself gardening was the treasured subsidiary occupation that we got to when we had spare moments; if I had my time again I would do as my daughter did and get some horticultural qualifications, do some professional learning and be more knowledgeable.

We made plenty of mistakes. Not that that is a disaster. In gardening, like anything else, you learn your own taste, you discover what you like and do not like, by way of trial and error. We learned that we liked profusion, variety, clematis climbing up old apple trees, snowdrops and leucojums, roses, roses all the way . . . And much, much else. It occurs to me that discovery of gardening taste has a parallel with discovery of reading taste: as you read, growing up, in adult life, you discover the sort of writing you want, and as you garden you find the plants you want, the way you want them to look, the image of the garden you are trying to create. I have moved from garden to garden, the two streams and the prolific vegetable growing are decades ago now, and I have ended up with a small paved London back garden. Which means that a new garden taste has to be identified and developed. No daisy-sprinkled lawns, no yew hedges – I am a pot specialist now. Fuchsias, heuchera, hostas, plenty of geraniums in the summer, tulips and various narcissi in the spring. Hydrangeas – I have become adept at finding which hydrangeas will do nicely in a large pot. And with limited bed space roses have to be very carefully chosen – two David Austin climbers up the wall, low-growing ground-cover roses at their feet. Snowdrops and grape hyacinths and 'Tête-à-Tête' daffodils in front, and a select corner for hellebores. The climbing hydrangea up the back wall, and *Hydrangea paniculata* 'Limelight' tucked into the corner bed. And anything that takes my fancy in pots that can be moved around, so that the garden changes week by week, month by month. The smallest of gardens can be made to perform, to mutate from season to season. I miss the long-gone days of digging a trench for the potatoes, pruning a bed of roses, dividing the irises, but there is immense satisfaction in the intimacy of a restricted area, where no space can be wasted, everything has to be considered, cherished, made to do its best.

Most of us garden according to the dictates of the day. Garden fashion. It was ever thus. There are always gardening pundits, those whose expertise and talent will inspire everyone else. In the early part of the twentieth century William Robinson and Gertrude Jekyll swept aside the Victorian passion for carpet bedding – thousands of annuals laid out in colour formation and patterns – and showed how to garden with the emphasis on structure and harmony, a natural look achieved in fact with subtle emphasis. Robinson showed how to plant drifts of narcissi and other spring bulbs in sweeps of grass. Jekyll favoured silvers and blues in her carefully constructed borders, *Erigeron karvinskianus* tumbling from high stone walls, marrying planting with landscaping when she teamed up with Edwin Lutyens: rill gardens – a canal lined with irises, sunken paved rose gardens with curved stone seats. We still tend to garden according to Jekyll, but the later part of the century brought other pundits, other tastes. And, crucially, the influence of television gardening programmes, which have had the nation rushing to create a water feature, install wall-to-wall decking, try to tuck a meadow into the sparse territory of a suburban semi. And garden centres. I have to admit here that I am a pushover where garden centres are concerned, unable to resist some choice new offering, loading the trolley with yet another grass, fern, or tempting plant I haven't had before. I am more phlegmatic where the television programmes are concerned, appreciative often, but also irritated occasionally by presenter style, fashion decrees that you know will at once influence what the garden centres will stock. That said, now that I can no longer do much, if any, garden visiting of my own I do often relish that window into delights that I shall never see.

Time was, the time of the garden with two streams, we were part of the Yellow Book garden openings – people whose gardens are open to visitors under the

National Garden Scheme. We only just scraped in, I think – the Garden Scheme inspectors are steely-eyed and have high standards – but on one Sunday every summer many interested and beautifully behaved people would cruise through the garden, and by the end of the day we'd have contributed to the large sums of money that the scheme raises for charity. We did a great deal of Yellow Book garden visiting ourselves. There is no better way to discover how other people garden, to get ideas, admire, fail to admire. I can remember some revelations: the National Collection of corokias in a north London garden, an amazing assembly of auriculas in someone's tiny backyard.

The corokias interested me particularly because I have one myself – and my respects here to any reader who knows what a corokia is. They are shrubs or small trees, a species native to New Zealand, and attractive for their twisting grey stems and light foliage – mine has been living in a large pot in the middle of the London back garden for nearly thirty years, and I like it for its see-through quality, and, now, its longevity. It gives height in the centre, but does not block the view of the Japanese tassel ferns in two white pots at the back of the garden. Most gardeners become selective in their plant interests. You can't like everything – indeed most of us acquire strong dislikes – and most of us have favourites. I remember my grandmother saying thoughtfully that if she ever ended up with what she called 'a pocket handkerchief garden' (comparing our suburban semi to her Somerset acreage) she would grow just one thing, to perfection. 'Probably irises,' she reflected. And I can see her point – it is that of a professionally minded gardener.

I am addicted to fuchsias, violas, hydrangeas, *Erigeron karvinskianus* and so on. It is no good lusting much after roses in a small back garden, so the choice has to be frugal. I realise that I don't much miss grass, a lawn. A passion for grass cutting and possession of a stable of lawnmowers is an entirely male condition, in my view. I miss trees. The garden with two streams had two lovely silver birches. And a quince – I am nostalgic for the smell of a bowl of ripe quinces on the kitchen table. And another Oxfordshire garden had a *Quercus ilex*, the holm oak, a Mediterranean tree that has bravely established itself here, with a *Robinia pseudoacacia*, or false acacia, next to it, another non-native, from North America this time. The dark shiny leaves of the holm oak contrasted nicely with the light, golden-green foliage of the robinia, and I rather liked the immigrant status of both trees.

I live today on a garden square, and relish it for its trees – some rare surviving elms, a chestnut, a laburnum, lilacs, flowering cherries. London is rich in garden squares; some are those snobby ones open only to residents of the square, but mine is council property, open to all. This means of course that it is open also to abuse – we have the occasional spell of drug dealing, or illicit barbecue parties – but on the whole the space is respected, and most frequented by mothers with small children, the elderly enjoying a sit in the sun on a bench. During the coronavirus lockdown, I exercised there every day, walking circuits of the garden's outer path, and always noting the birdlife: blackbirds, of course, great and blue tits, dunnocks, a pied wagtail one day, a jay on another, the rapacious pigeons, crows, magpies. The garden is home to many (a rough sleeper, indeed, for a while, tucked under a hedge in a sleeping bag until the park warden paid a rare visit and chivvied him out), a precious space in the city, a kind of oasis of green and growth amid the tarmac and brick, roads, buildings, urban conglomeration.

Public gardens, parks, are essential lungs, a breathing space for everyone. Not everyone can have a garden, some people don't want a garden, but everyone needs a park at one time or another. London is rich in parks, both large and small – it is always a pleasure to look at the map and see it splodged with green here, there and everywhere. And trees . . . So many tree-lined streets. One has to admire in particular the London plane, *Platanus x acerifolia*, so robust that it seems able to endure any amount of urban pollution, and handsome too with its fine trunk and sprays of black seed pods in winter. There are some inner-city squares with planes a hundred years old and more, weathering two world wars and decades of seismic social change, presiding over it all, resilient and impervious.

If you are a gardener, you have an extra eye – the gardening eye. You always notice what is growing when out and about. These days, in my old-age London days, I notice window boxes – someone who has bothered with an effective summer combination of white petunias, trailing bacopa, blue lobelia, another person's winter display of pink cyclamen. Central London in summer is resplendent with what I think of as the professional window box, those supplied by companies, a riot of begonias, geraniums and everything else – all very fine, but I prefer those that are more homely and personal, that someone has thought about and chosen. My gardening eye always takes note too of basements – those spaces below the pavement that someone, usually a basement-flat dweller, has laid hands on and filled up with a collection of cacti, a resplendent flowering jasmine, a bamboo, a fatsia, a stack of potted geraniums. London's acreage of early nineteenth-century terrace houses provide a wealth of basement space – look down, as you walk, and you will find the subterranean gardens.

In my country days, the gardening eye was differently focused, and was looking beyond gardens, into the landscape, at what grows anyway, ungardened, unkempt. Wild growth – the cow parsley, red campion, primroses, meadowsweet, bush vetch, toadflax, scabious, scarlet pimpernel, yarrow, herb-Robert, everything that springs up everywhere unbidden and untended. Weeds, when they decide to invade the garden. But many of them, if rare and requiring cultivation, would be sought-after items in garden centres – we would be loading up our trolleys with cow parsley and toadflax. I have always needed to know the name of whatever it is I am looking at out there, my ancient copy of Keble Martin (*The Concise British Flora in Colour*) is annotated on page after page, what was found where and when.

The gardening eye, the eye that notices what grows – you either have it or you don't. If you don't, you perhaps have an eye for other things that we gardeners are missing. But gardeners are a community, a huge community with many sects, from those like me today with just a small patch at the back of the house, to those with rolling acres, and those who do it professionally – designing gardens, telling the rest of us how to do it – or simply growing the very best hostas or clematis or roses or primulas or peonies. Once a gardener, you find yourself offered a whole new dimension of experience, and an insight into the world of things that grow. You not only want to have a go at it yourself, whether you have a window box, a few square yards of basement space, or one of the substantial suburban gardens up and down the land, but you also now have a permanent interest and curiosity – you want to see more, know more. You will become a garden visitor, grace of the Yellow Book, the National Trust, the Royal Horticultural Society; you will learn from, or quarrel with, the television gardening pundits; and, most of all, you will be forever needing to find out what that unfamiliar plant is called. Where did it come from? Do I want to grow it?

Coming of Age

NIGEL SLATER

I could barely wait to dig up the lawn. Getting rid of the rectangle of mown grass that passed for a garden was almost the first thing I did when I moved into my new home on a bitterly cold New Year's Day, 20 years ago. An act of vandalism to some, but to me the patch of brown earth, chips of broken china and old, soil-filled medicine bottles it left behind was just the blank canvas I needed.

The wielding of the spade must have been symbolic, because I did nothing more outside for months. I wanted to create a place to think. A green space in which to clear my head in between recipes, or to untangle a knotted sentence. A lawn has its uses as a safe place for children to play, somewhere to kick a ball around or to sunbathe, but it wasn't the right garden for me. I don't find inspiration or peace in a neatly edged rectangle of grass. All I see is a wasted opportunity. I wanted a garden where my imagination could run, where I could make a home for bees, birds and butterflies and where I could escape to when the house was full of people. A place that would act as both inspiration and sanctuary.

It was then that Monty Don, at the time the *Observer*'s gardening correspondent, came to lunch. Over deep bowls of pumpkin soup and homemade oat bread Monty hatched a plan, drawn in black pen on the back of an envelope. It was a plan that opened my eyes to the possibilities even the smallest urban space held. Monty's drawing couldn't have been further removed from a lawn – and I implemented it to the letter.

That brave new garden, with its smart box-edged vegetable beds, rows of peas and beetroot, hazel wigwams of beans and assortment of berry bushes, brought with it a decade of unimaginable joy. Heritage carrots were munched within minutes of being pulled from the soil; hedges that framed rows of cabbages and kale were clipped into soldierly neatness and purple-podded beans wound their way up cane frames. The garden went on to inspire two books (*Tender: Volumes I* and *II*), and a television series, and instilled an everlasting connection between growing, cooking and eating.

And then the new neighbours moved in.

At first I welcomed the fresh arrivals to the terrace, with their cute red-haired kids, and even tolerated their occasional antisocial behaviour. But then we fell out. I cannot exaggerate the amount of damage a large family of foxes can do to an urban garden if they are so minded. Every morning I woke to a new scene of devastation: rows of parsnips dug up and abandoned, lupins sat upon and pristine hedges crushed where they had been used as trampolines by the cute little cubs. Oh the delight of finding the contents of your neighbours' bin bags scattered over your pumpkin patch, not to mention pizza boxes, nappies and endless half-chewed trainers. I once found someone's old pants in the rhubarb and a spooky, dismembered doll in the fennel. And don't even start me on the subject of fox poo.

Each year the damage got worse, and every early-morning discovery of flattened tomato seedlings and abandoned takeaways was more heartbreaking than the last. To add to the fun, a plague of box blight (*Cylindrocladium buxicola*) arrived to denude my precious hedges of their leaves, thus destroying the backbone of the garden. What had been a place of inspiration and delight was now one of frustration and heartache.

It was then that I met the garden designer and writer Dan Pearson, who wielded his magic over the long section beyond the vegetable garden, turning what had previously been a wilderness into a magical space of delicate and thoughtful planting.

I inserted yew hedges to turn the long, thin patch from one garden into three very separate but homogeneous spaces. In came white *Cornus kousa* and quivering yellow epimediums; white hydrangeas were underplanted with woodruff; climbing roses tumbled among orange blossom. This meant you walked from the vegetable beds through a yew hedge into a fragrant, almost woodland space. The foxes departed for pastures new and, once again, the veg-growing started in earnest. I had my inspiration, my retreat and sanctuary.

It was then that a relatively new invasion came to visit. *Cydalima perspectalis*, the box tree moth, had arrived. In the space of a fortnight, the little horrors chomped their way through hundreds, no thousands of pounds' worth of topiary. Finding this one setback too many – and, I suppose, listening to reason – I realised that the vegetable garden in its present form had to go.

One afternoon in late spring I asked Katie, who has helped me in the garden for years, to rip up the denuded hedges. Those long lines of crisp-edged *Buxus* were reduced to a skipload of powdery grey twigs. It felt like a bereavement. The day they went I almost cried. Not for the first time, I was left with a blank canvas.

I learned quite quickly that every disaster in the garden is an opportunity in disguise. Dan's garden, as we call the woodland-inspired middle garden to this day, continued to establish itself, and left me free to rethink the space that housed the old veg beds. I had always worried that the space was too tightly packed. There was no place to eat or even to sit. It was a garden you walked through rather than lingered in. Putting pen to paper once again, I knew immediately I needed a table at which to eat and work. Somewhere for the neighbours' cats to curl up and sleep without crushing a courgette, and where you could just sit and breathe. There must also be space for my three great garden loves: topiary (though obviously not of box), ferns and climbing roses.

And so the garden moved into its third and probably final phase. The old vegetable beds took on a new role as a place to eat, surrounded on both sides with borders thick with ferns, waving white Japanese anemones and clipped topiary. The kitchen walls would now spend their summers shrouded in white wisteria and climbing roses. In summer the back of my house could relax into a tangle of carnival-coloured dahlias and trailing jasmine. There would be room for dahlias and dinner.

Of late, the garden has settled into a gentle rhythm. Once a year, on a dry spring day shortly after the

Chelsea Flower Show, everything gets a serious trim – the 'Chelsea chop' as it is known. Hedges are clipped, topiary is shaped and overhanging branches of the fig and medlar tree are pruned. A tidy-up that might appeal to the sort of gardener who power-washes their flagstones and scrubs the moss from their pots but, to me, it feels as if a much-loved and elegantly ageing friend has gone in for a round of cosmetic surgery. Not unrecognisable, but slightly cold and distant and, to my mind, a little dishonest. For a couple of weeks a year the garden doesn't quite feel like mine.

As autumn approaches, your way along the neat gravel paths of old is now delightfully hampered by drooping branches and heart-shaped yellow and purple leaves that brush against you as you pass. You push your way through collapsed magenta and orange dahlias and try not to slip on the figs that lie splattered over the terrace. You have to beat a path to get from one end to the other and the three distinct spaces relax into one.

I would like to say that the garden I have now will probably be my last. Twenty years on from digging up the lawn, I have a space that is more inspirational and restful than I could have ever imagined. I feel the garden has come of age.

Yet the space still refuses to stand still. Even now there are changes afoot. This year I reintroduced the vegetables and sweet peas that I missed so much. Tomatoes and calendulas now grow in huge terracotta pots on the kitchen steps and there is an entire table of culinary herbs. There are tubs of marigolds and stands of bronze fennel. Next year there may be more. The garden will never be 'finished'. I have no idea of what will happen next. All I know is that there won't ever be a lawn.

The Garden I Have in Mind

JAMAICA KINCAID

I know gardeners well (or at least I think I do, for I am a gardener, too, but I experience gardening as an act of utter futility). I know their fickleness, I know their weakness for wanting in their own gardens the thing they have never seen before, or never possessed before, or saw in a garden (their friends'), something which they do not have and would like to have (though what they really like and envy – and especially that, envy – is the entire garden they are seeing, but as a disguise they focus on just one thing: the Mexican poppies, the giant butter burr, the extremely plump blooms of white, purple, black, pink, green, or the hellebores emerging from the cold, damp, and brown earth).

I would not be surprised if every gardener I asked had something definite that he or she liked or envied. Gardeners always have something they like intensely and in particular, right at the moment you engage them in the reality of the borders they cultivate, the space in the garden they occupy; at any moment, they like in particular this, or they like in particular that, nothing in front of them (that is, in the borders they cultivate, the space in the garden they occupy) is repulsive and fills them with hatred, or this thing would not be in front of them. They only love, and they only love in the moment; when the moment has passed, they love the memory of the moment, they love the memory of that particular plant of that particular bloom, but the plant of the bloom itself they have moved on from, they have left it behind for something else, something new, especially something from far away, and from so far away, a place where they will never live (occupy, cultivate; the Himalayas, just for an example).

Of all the benefits that come from having endured childhood (for it is something to which we must submit, no matter how beautiful we find it, no matter how enjoyable it has been), certainly among them will be the garden and the desire to be involved with gardening. A gardener's grandmother will have grown such and such a rose, and the smell of that rose at dusk (for flowers always seem to be most fragrant at the end of the day, as if that, smelling, was the last thing to do before going to sleep), when the gardener was a child and walking in the grandmother's footsteps as she went about her business in her garden – the memory of that smell of the rose combined with the memory of that smell of the grandmother's skirt will forever inform and influence the life of the gardener, inside or outside the garden itself. And so in a conversation with such a person (a gardener), a sentence, a thought that goes something like this – 'You know, when I was such and such an age, I went to the market for a reason that is no longer of any particular interest to me, but it was there I saw for the first time something that I have never and can never forget' – floats out into the clear air, and the person from whom these words or this thought emanates is standing in front of you all bare and trembly, full of feeling, full of memory. Memory is a gardener's real palette; memory as it summons up the past, memory as it shapes the present, memory as it dictates the future.

I have never been able to grow *Meconopsis betonicifolia* with success (it sits there, a green rosette of leaves looking at me, with no bloom. I look back at it myself, without a pleasing countenance), but the picture of it that I have in my mind, a picture made up of memory (I saw it some time ago), a picture made up of 'to come' (the future, which is the opposite of remembering), is so intense that whatever happens between me and this plant will never satisfy the picture I have of it (the past remembered, the past to come). I first saw it (*Meconopsis betonicifolia*) in Wayne Winterrowd's garden (a garden he shares with that other garden eminence Joe Eck), and I shall never see this plant (in flower or not, in the wild or cultivated) again without thinking of him (of them, really – he and Joe Eck) and saying to myself, It shall never look quite like this (the way I saw it in their garden), for in their garden it was itself and beyond comparison (whatever that amounts to right now, whatever that might ultimately turn out to be), and I will always want it to look that way, growing comfortably in the mountains of Vermont, so far away from the place to which it is endemic, so far away from the place in which it was natural, unnoticed, and so going about its own peculiar ways of perpetuating itself (perennial, biannual, monocarpic, or not).

I first came to the garden with practicality in mind, a real beginning that would lead to a real end: where to get this, how to grow that. Where to get this was always nearby, a nursery was never too far away; how to grow that led me to acquire volume upon volume, books all with the same advice (likes shade, does not tolerate lime, needs staking), but in the end I came to know how to grow the things I like to grow through looking – at other people's gardens. I imagine they acquired knowledge of such things in much the same way – looking and looking at somebody else's garden.

But we who covet our neighbour's garden must finally return to our own, with all its ups and downs, its disappointments, its rewards. We come to it with a blindness, plus a jumble of feelings that mere language (as far as I can see) seems inadequate to express, to define an attachment that is so ordinary: a plant loved especially for something endemic to it (it cannot help its situation: it loves the wet, it loves the dry, it reminds the person seeing it of a wave or a waterfall or some event that contains so personal an experience as when my mother would not allow me to do something I particularly wanted to do and in my misery I noticed that the frangipani tree was in bloom).

I shall never have the garden I have in my mind, but that for me is the joy of it; certain things can never be realised and so all the more reason to attempt them. A garden, no matter how good it is, must never completely satisfy. The world as we know it, after all, began in a very good garden, a completely satisfying garden – Paradise – but after a while the owner and the occupants wanted more.

The Collective Garden

A Common Inheritance

FRANCESCA WADE

In his *Tribune* column of 4 August 1944, George Orwell noted that the railings – removed at the outbreak of war for scrap iron – were returning to London squares, so that 'the lawful denizens of the squares can make use of their treasured keys again, and the children of the poor can be kept out'. Their destruction, he argued, had served as a 'democratic gesture': the gardens had acquired a 'friendly, almost rural look', since visitors were no longer 'hounded out at closing times by grim-faced keepers', but could enjoy the green spaces at leisure. The article attracted lively debate, including one stiff response which suggested that Orwell, in proposing that square gardens remain open to the public in perpetuity, was advocating the theft of private property. Orwell replied:

> If giving the land of England back to the people of England is theft, I am quite happy to call it theft [. . .] Except for the few surviving commons, the high roads, the lands of the National Trust, a certain number of parks, and the sea shore below high-tide mark, every square inch of England is 'owned' by a few thousand families. These people are just about as useful as so many tapeworms [. . .] For three years or so the squares lay open, and their sacred turf was trodden by the feet of working-class children, a sight to make dividend-drawers gnash their false teeth. If that is theft, all I can say is, so much the better for theft.[1]

The story of London's garden squares is full of such disputes. For centuries, their railed enclosures have stood as symbols of both the best and worst of the city: the charm and satisfaction of stumbling upon a tranquil green oasis in the midst of polluted streets, all too often tempered by the frustration of finding its iron gates locked, and having to settle on the kerb to eat your sandwich while just beyond reach a multitude of benches rest empty. The garden square's fraught history of boundaries, hierarchy and exclusion encompasses immense social changes – to class politics, to the way we conceive of the family, to notions of domestic privacy – as well as providing insights into the subtleties of sewage systems, shrubbery lifespans and building contracts. It's a history of architecture, of protest, of literature; a conversation down the generations between town planners, commentators and individual citizens, all proposing speculative answers to the question of how Londoners might want to live. It starts as a story of monarchs and aristocrats – of lands transferred between families, of enterprising earls and determined dukes, seeking to capitalise on their holdings by parcelling out land for residential developments. But as the story of garden squares continues, it offers a microcosm of wider, ongoing shifts in the fabric of the city. When the railings were torn down, disgruntled landowners asked over and over with feigned humility, 'Of what benefit to the public are any of our dingy, uninhabited squares?' Yet the pleasure of exploring these small urban gardens – whether square, circle, crescent or polygon – endures; no walk around London is complete without gazing through bars or gaps in hedges at ancient trees and elegant fountains, imagining the stories hidden within.

The London square was born in 1631, when the Bedford family began to build Covent Garden on pastureland previously let out for sheep to graze, commissioning a design by Inigo Jones which took inspiration from the grand piazzas of Italy. Lincoln's Inn Fields (1638) and Southampton (now Bloomsbury) Square (1661) were among the first to follow suit, as the ancient city began to expand and its discrete pockets to connect and coalesce. From this point, new squares, generally featuring ornamental mansions surrounding an open space, began to spring up apace across London and its new suburbs: some, like Queen Square in Bloomsbury, were built on the outskirts of town to take advantage of resplendent views out to the distant countryside of Hampstead and Highgate; others, like those in Hoxton and Kensington, extended the city's boundaries to the east and west. Contracts for these new developments regularly stipulated that the central space should be retained as a garden, often incorporating the ancient trees that stood on the former fields. Creating the ideal of *rus in urbe* – a small sliver of wilderness inside the city walls – was a process of taming, cultivating, refining: transforming open pastureland into a garden enclave that might become a haven for wild birds and exotic plants, nestled within the bustling city.

By the middle of the eighteenth century, squares were London's defining architectural feature, admired across Europe not only for their grand plazas but for the smaller lanes, courts and mews that surrounded them, creating a constellation of dwellings with distinct character. As London developed as a commercial centre, many small squares were built on the former gardens of sprawling mansions, providing homes for professional workers eager to enjoy a countrified atmosphere within easy reach of the office. Trollope and Thackeray's characters, attuned to the vagaries of social status, debate the merits of new squares in Pimlico against the traditional glamour of Belgravia; Bloomsbury, now the district most synonymous with garden squares, then remained a murky middle-ground, poised in the minds of the upwardly mobile between respectability and danger. The fortunes of neighbourhoods could change as quickly as gardens overgrow and houses fall into disrepair. In 1630, Robert Sidney took a lease on the meadowland of Leicester Fields (promising that his building plans would leave 'fitt spaces [. . .] for the Inhabitantes to drye their Clothes there as they were wont').[2] By 1670, he had begun to develop terraced houses around the edge of the fields and plant the centre with flower beds and elm trees, creating the first iteration of what is now Leicester Square. Two centuries later, the land had been sold on and redeveloped several times, with new streets – and the growth of London's entertainment district – leaving the area packed with previously unimagined traffic and trade. The journalist George Augustus Sala described Leicester Square's garden as 'a feculent, colourless vegetation like mildewed thatch upon a half-burnt cottage', a 'howling desert enclosed by iron railings', where the shrubbery was infested with soot-drenched weeds and oxidised saucepan lids, and wild cats as large as leopards slept in the thickets.[3] In 1874, the land was turned over to the Metropolitan Board of Works to create a new 'people's garden', complete with a children's playground, and as a centrepiece a fountain featuring a statue of William Shakespeare surrounded by frolicking dolphins.

The democratising move was greeted by some with suspicion, by others with joy. Social reformer Octavia Hill seized upon the developments as a precedent for other 'sternly secluded' city gardens to transform into 'people's pleasure-grounds', open without restriction to children living in cramped and overcrowded tenements across the city. Hill – primarily known for her work on social housing – was one of the founders of the National Trust, which campaigned against threats to Hampstead Heath and other free open spaces in London, and sought to shield London's disused burial grounds and decrepit garden squares from the gleaming eyes of developers. Her mission was 'to protect all open spaces, no matter how small, secure them, cultivate them, and most importantly allow them to be open "for everyone"': to her, London's garden squares were a 'common inheritance', which should be shared between neighbours and provide every person, regardless of wealth or background, with 'places to sit in, places to play in, places to stroll in, and places to spend a day in'.[4] This, Hill insisted, was not a luxury, but an essential.

Hill's ideas gathered momentum through the nineteenth and twentieth centuries. Over the late Victorian era many once-fashionable squares were deserted, while newer ones languished uninhabited, as the wealthy residents for whom the grand houses had been built moved to serviced mansion-flats or further out into the suburbs where they might enjoy the luxury of their own private garden. As speculative builders circled around abandoned or unkempt garden squares, lamenting these strips of open space as a waste of valuable real estate, Hill launched a campaign for 'London Gardens for the Poor', demanding that these spaces be handed over to local authorities. Debates over the preservation of railings, such as Orwell would reanimate in the 1940s, were a regular feature in the press: a correspondent to the *Sanitary Record* in 1878 declared that 'a square without railings would always be under the eye of the police', while in 1874 the *Graphic* attempted to raise sympathy for the Londoner who had chosen the peace of a private garden square: 'there is nothing more annoying to the ears than the hum of multitudes and the crooning of children'. But if garden squares were built as organs of class segregation – with gardeners regularly doubling as security guards – it was clear that they could also become the symbols of a more equal, democratic and healthy city. Increasingly, councils stepped in to buy square gardens threatened with redevelopment; negotiations with the landlords of Edwardes Square in Kensington led, in 1906, to the passing of the London Squares and Enclosures (Preservation) Act, which ruled that squares were part of the city's essential character. In 1931, it was shored up by more robust legislation designed to conserve the spaces known fondly as 'the lungs of the city'.

'I like this London life in early summer,' wrote Virginia Woolf in her diary in April 1925, 'the street sauntering and square haunting.' Square haunting is free, nourishing, imaginative and invigorating: it can be done on a whim, in a lunch break, in summer (when the gardens are in full bloom) or winter (glancing in through windows at people's Christmas decorations), in a spirit of determination or meandering, tracing a map or just following the most alluring turns as they loom ahead or emerge unseen on either side. It can be done in almost any corner of London, too. There are more than 300 garden squares across the capital, each with their own shapes and characters: Dickens, a great London street-saunterer, noted the 'aristocratic gravity' of Mayfair's finest squares, the 'dowager barrenness and frigidity' of those in Fitzrovia, and the draw of those deserted inner-city spots 'favourable to melancholy and contemplation'.[5]

Bloomsbury, of course, is the place to start: with Gordon Square, peppered with blue plaques to the group who (in words attributed to Dorothy Parker) 'lived in squares, painted in circles and loved in triangles'; Tavistock Square, with its statue of Mahatma Gandhi, bust of Virginia Woolf and commemorative plaque for conscientious objectors from all nations and eras; Bedford Square, still the estate's crown jewel and the first London square since Covent Garden to be built to a uniform design; and Russell Square, once famous for its flower shows where local residents would display the plants they had grown on window-sills and balconies. But wander out to east London to admire the historic bandstand in Arnold Circus, the centrepiece of one of the city's first social housing estates; majestic Tredegar Square in Mile End (opened to the public in 1931 by Prime Minister Clement Attlee); or Bartholomew Square by Old Street round-about, London's first floodlit playground. Northwards in Islington or westwards around Kensington, squares unfold like a chessboard: Canonbury, Arundel, Gibson, Alwyne, Lonsdale; Onslow, Cadogan, Lowndes, Eaton, Belgrave. Down south to Camberwell, there's

Addington Square, developed along with the Grand Surrey Canal, which once boasted a private bathhouse on its north side, and the former Brunswick Square, a never-completed development of detached homes, which opened as a public park in 1907, leaving two houses tucked awkwardly inside the modern garden. Wending back to Bloomsbury in the gloaming, the gardens turn over to prowling foxes; curtains are drawn and the ancient plane trees stand silhouetted against the night sky.

As square living fell into disrepute during the late nineteenth and early twentieth centuries, they tended to pass to writers and artists whose homes and living arrangements often became synonymous with their radical visions: from Jean Rhys, whose heroines often scrape by to pay the rent on seedy boarding houses in run-down Bloomsbury squares, to Ted Hughes and Sylvia Plath's long occupancy of a tiny flat at 3 Chalcot Square in Camden Town; from the congregation of the Pre-Raphaelite Brotherhood around Red Lion Square (where in the 1860s William Morris and Edward Burne-Jones ran their Decorators' Shop, producing affordable but enduring furniture and textiles in protest against the mechanisation of manufacture), to Roger Fry's Omega Workshops, which operated from 33 Fitzroy Square some half-century later on similar egalitarian principles (with Vanessa Bell and Duncan Grant among its collaborators). The spirit of both outlets – founded on the idea that beauty and good design should be available to all – encapsulates something of the London garden square itself, when the railings are pulled down and the gates flung open. These are places of the imagination: containers not only of historical or literary memory, but of personal pasts too. Each of us charts the city by our own haunts, movements and memories, imprinting these public spaces with the private vestiges of our own experience. London's city gardens are places of sanctuary: to gather for picnics or protests or to be alone, and they belong to us collectively. As rents in the capital grow and green space becomes ever more scarce, public space continues to be privatised by a new set of landlords, the aristocrats of old replaced by big business, corporations and private developers who buy up land with few obligations to the people who live on or care for it. In the meantime, though, these gardens are ours to lay claim to, tend and enjoy.

Companion Planting

JON DAY

We started the garden by mistake. A few years ago my partner Natalya and I moved from the centre of town to the suburbs of east London. For the first time in our adult lives, we had a garden. It was a dark, north-facing plot, overshadowed by neighbouring trees and houses. There was a scrubby lawn. The borders were littered with dried cat shit and the chewed cow bones our neighbours fed to the foxes each night. We had moved in winter, and as spring arrived a clump of Japanese knot-weed at the back of the garden burst into life, thrusting its spear-shaped leaves up through a pile of rubble. Not knowing what it was, we thought it looked quite pretty.

For a long time we neglected the garden. But a few years after we'd moved, the autumn after our daughter Dora was born, we began to clear the dead branches and rubbish. Newly homebound, we wanted space: somewhere for us to sit outside, where Dora could learn to crawl. We cut back overgrown shrubs, laid a path, built raised beds and doused the knotweed with glyphosate, worrying that we would never be rid of it, or that we'd poison ourselves in the process. We still didn't really know what we were doing, but soon we developed a vague, inextricable urge to watch *Gardener's World*. At weekends I found myself eyeing up hedge trimmers in B&Q.

Across the road from our house there was a cricket pitch, the old Essex County Ground. It had a handsome, dilapidated pavilion in which pigeons now roosted, and was surrounded by a high concrete fence topped with barbed wire. At the northern edge was a small patch of wasteland, which had once been a bowls club. The club had closed a decade ago, and the bowling green was now covered in weeds. The clubhouse had long since been demolished.

Just as Natalya and I were getting to grips with our own garden, the council announced plans to build a playground on the site of the old bowls club. They wanted to include some beds in the design, they said, to allow local people to grow vegetables and flowers, to learn about food and commune with the earth. There was vague talk of helping us get our five a day, and a slogan: 'living better; living healthier'. When the building work on the park was completed the Mayor of Leyton planted a sorry-looking rosemary bush in the sandy soil and declared the beds open.

The following Saturday a few residents met in the park, where Kate, a friendly woman from the council's 'green spaces team', taught us about germination and soil structure and showed us how to sow seeds. There were only six of us there that day, but in the weeks that followed more people joined, like moths drawn to a light. People would come over to see what we were doing, and soon they'd have borrowed a spade and started weeding. Others would sit and watch, while children played in the park or ran around their legs. Talk was easy: focused on the task in hand, any silences were never awkward.

After a few months there were a dozen or so of us looking after the beds. We were a motley crew: middle-class parents, like us, who had only recently moved to the area; older locals, often first-generation immigrants from Bangladesh or the West Indies, whose children had grown up and moved away; young families from eastern Europe. People brought their knowledge to the garden, but they also brought their plants: plants that reminded them of home, or that they couldn't easily buy in local supermarkets. Mohammed, a waiter who'd lived in the area for thirty years, planted lablab, a beautiful, purple-flowered bean, which is poisonous if not prepared correctly. 'It's £7.50 a kilo in the market,' he said. Una planted a variety of Jamaican squash from seeds she'd brought back from a visit home a few years ago. Over the summer it took over an entire bed, its tendrils bursting with bulbous fruits as autumn came.

One of the techniques Kate taught us was companion planting. Companion plants are plants that grow well together by shielding each other from the elements, offering sacrificial protection from pests, or enriching the soil by trapping nutrients. Often plants that grow well together taste good together: basil and tomatoes thrive in similar conditions, and grow in sympathy with each other too. Before the tomatoes get too tall, the basil can flourish between the rows. Nasturtiums often thrive when sown among other plants, protecting them from blackfly and covering the ground to prevent weeds from becoming established. Probably the most well-known, and certainly the most venerable, companion plants are the 'three sisters' – squashes, beans and corn – which have been grown together by subsistence farmers in South America for thousands of years. The corn provides a structure for the beans to climb, and the beans fix nitrogen into the soil to feed the squashes. Companion plants are often very unlike in their pairings, but they go to build a community, which is what a garden is.

That first year we mainly grew vegetables – tomatoes, broccoli, beans – as well as herbs and flowers. So much of the joy of gardening seemed to be about the promise of it. Going through seed catalogues and picking bulbs felt like an investment in the future. We became obsessive composters – orchestrating the delicate ratio of plant matter, air and water, which, if you get it right, turns magically into a rich, life-sustaining mulch. By the end of the summer our sunflowers towered over our heads as we worked. We marvelled at the beauty of the beds and the abundance of our harvest – knobbly Jerusalem artichokes, armfuls of vibrant chard, buckets of potatoes and carrots – and began looking for more space in which to grow.

Over the following year our group expanded. There were now a hundred people subscribed to the email list I used to plan our activities. Each week thirty or so would turn up in the park to garden. But we were running out of space, and so, a year after we first started the garden, I wrote to the council to ask if we could build a greenhouse on the patch of land next to the park, down the centre of which ran an old access road to the cricket ground. Our needs were modest. We wanted a place to store tools, and a greenhouse in which we could germinate seedlings before planting them out in the beds in the park. Spotting an opportunity for free maintenance, the council seized it: why didn't we take on the whole site, they suggested. We would have to keep the place tidy and maintain the access road, but other than that we could do what we liked.

Though the new space – a narrow quarter acre of tarmac and thick concrete, with brambles and bindweed spilling out through the cracks – wasn't huge, the scale of it was still daunting. The ground was bare and hard, and it was difficult to see how we might get down to the soil underneath, or what we'd find when we did. But it was an open plot, and had a beautiful, high wall running along its eastern boundary, which trapped the sun's heat during the day and radiated it out again in the evenings.

We got to work, building raised beds from railway sleepers and filling them with dusty topsoil. We dug in compost, filled with all manner of domestic waste – broken bones, bits of plastic and old batteries – which we sourced from the council tip. We found a second-hand greenhouse on a local Facebook group and patched it up with plastic panes, zip-ties and string. We borrowed a van and drove to some riding stables deep in Epping Forest, where we filled dozens of bags of manure to condition the soil, providing nutrients and the colonies of insects and microbes which would give it life.

The Boundary Garden, as it became known (a nod to the cricketing history of the site, and to the narrow shape of the plot), began to flourish. We grew vegetables and fruit, but we also wanted to leave space for insects and birds. So we planted half the new beds with crops – potatoes, onions, soft fruit, asparagus – and half with flowers. In the largest bed, which ran along the length of the wall, we sowed a wild flower meadow: in May it burst into an impressionist haze of opium poppies and cornflowers, turning orange as swathes of *Coreopsis tinctoria*, a dyer's plant, flowered in July. The meadow became a refuge for insects, and soon birds came too: goldfinches flew in to eat the dried seed heads, wagtails nodded along the edge of the tarmac.

We built a polytunnel so that we could grow more vegetables over winter and applied for a grant from the Mayor of London, which we used to pay for a digger to smash up some of the tarmac and concrete to create space for more raised beds. By the end of the first year the new garden was thriving. We didn't have many rules, but the Boundary Garden had some founding principles. The first was that no one should pay for anything: we became good at finding small pots of corporate and council funding. The second was that no one would have their own individual plot in the garden. We wanted to grow together, as one, and to share what we grew when it was ready.

We wondered whether it would be possible to sustain an urban garden on goodwill and trust. We thought our tools might be stolen, or that people would dig up the plants. We shouldn't have worried. These days we are open all the time. There is a gate, but we rarely lock it. Sometimes drinkers congregate there early in the morning, and young men gather in the evenings, filling the air with their cotton wool clouds of vape smoke. I once found someone sleeping in the greenhouse. There are traces of other activities too: reminders that more desperate people sometimes use the garden. Occasionally I find syringes in the gravel. Once I found a bullet case, lying brightly in the dust.

Perhaps it's just my age and stage of life (perhaps I just notice them more these days), but there seems to have been an explosion of interest in community gardening over the past decade or so. In my area of London every other street now hosts a pocket garden, or planters around each tree, which are tended by enthusiastic bands of volunteers. It's easy to read this engagement with vegetable life as an anxious response to the overwhelming political and environmental turmoil of the time: a growing awareness of the terrors of industrial farming and planetary extinction manifesting in a desire to eat local and live lightly. But as a political response gardening can feel like an indulgence: neither revolutionary nor progressive, but a distraction from other, more important struggles.

Gardens are often thought of as retreats, as private places isolated from the wider world, secret zones hidden behind high walls and forbidding fences. Perhaps this is one reason they are such rich sources of metaphor. In the Islamic and Christian traditions Paradise is a garden: a non-place devoid of work, from which it is possible to be exiled. But this separation also means that the activity of gardening can feel self-indulgent, as though looking down at the soil all the time prevents you from confronting the world at large. Rebecca Solnit has said that gardens are to millennials what sex and drugs were to the boomers who preceded them. 'Gardens are where they locate their idealism, their hope for a better world, and, more than hope, their realisation of it on the small scale of a few dozen rows of corn and tomatoes and kale.'[1] The political timidity of gardening, Solnit says, reflects a wider generational shift, away from direct action towards a commitment to the more nebulous categories of kindness and care. Gardens, she concludes, can be realms 'of quiet idealism', but that idealism can 'readily slide over into disengagement or the belief that your activism can stop with the demonstration of your own purity and lack of culpability'.

Can gardens be radical places, politically, socially, otherwise? Need they be? I've wondered about that a lot over the past few years, as the Boundary Garden has grown. Radical, of course, means 'to address something at the roots' (and therefore has the same root, etymologically speaking, as radish – we grow very good radishes in the garden). It is true that our garden, in a practical sense, is not going to save the world. It's certainly not an efficient way to produce food. Though we usually have a good annual harvest of onions and potatoes, we'd struggle to feed a single family on the produce we are able to grow.

But perhaps productivity, or political impact, are not the right measures here. Gardens are always mirrors of society. They can be exclusive, exclusionary places: manicured and maintained, with signs warning you not to pick anything or to keep off the grass. Or they can be messy, open, companionable endeavours, developing organically through the availability of resources – physical or material – of those who are able to contribute to them. As a child I loved Oscar Wilde's 'The Selfish Giant', a moral Christian fable about what it means to share a space with other people, and about what it takes to make a community. Gardens, as Wilde knew, might be apart from the world, but they are always already embedded in it. Gardens, like communities, are places. But they are processes and conversations too. And they only make sense if they're shared.

The Earth I Inherit

PAUL MENDEZ

Growing up as a Jehovah's Witness, my sense of the future was dominated by the image of a paradisal garden. The illustrated cover of the 1986 brochure *Look! I Am Making All Things New* was my vision of hope. A feast of bright colours, it was painted to catch the eyes of the poor, the young and the disenfranchised, depicting a world in which humans of all races lived in harmony with nature, and with each other. Under a blue cloudless sky, an African man, a Japanese woman (holding an apple), an Indigenous mother and her child and two European men laugh; they share fruits and sip water cut with fresh lemons. Behind them, a docile lion plays gently with a pure white lamb, right by a group of children frolicking around with their terrier as flamingos share a crystalline brook with swans. Mount Fuji is portrayed in the background, with traditional Japanese and modern American-influenced homes dotting the hillsides. Bushes and trees, English-looking flowers and tropical birds, chosen more for their colours and fan-tailed flamboyance than their verisimilitude, complete the visual offering. You can just hear the squeaking of parakeets, the call of kingfishers and the tinkling of water over rocks. Who would not want to live here?

This is the Witnesses' vision of a Paradise Earth: our entire planet rid of all evil, restored to (an artist's impression of) its natural perfection. It promises people that they can help rebuild and live forever in the idyllic conditions our common ancestors, Adam and Eve, gave up when they chose to defy Jehovah and were punished with exile into a thorny, thistly, inhospitable wilderness. My grandparents, all four of them members of the Windrush generation and born in Jamaica, must have felt, when they arrived in 1950s England, as Norman Alonso did in my novel *Rainbow Milk*: that they had left 'the Garden of Eden for the Land of Milk and Honey' only to find 'Sodom and Gomorrah'. I imagined Norman, with his wife Claudette, would have dreamt of the Mother Country as Hortense Joseph in Andrea Levy's *Small Island* did: of 'verandas', 'manners', 'politeness' and 'refinement', 'a cold morning with daffodils blooming with all the colours of the rainbow'.

My character Norman is a self-taught gardener who has worked for an Englishman in the grounds of a former plantation house. His two black children play with their blocks on a perfect green lawn. He fills his borders with bare roots from David Austin, a family-owned rose nursery that still thrives today outside Wolverhampton. Norman chooses cultivars for their fragrance, to prevent further complaints from his neighbours about the smell of Jamaican cooking. He plants jasmine at the back to combat the odours drifting up from the nearby gas works. His roses are strong and bountiful, the envy of the street. But jealous neighbours sneak up brandishing secateurs. His home is the victim of fascist graffiti. And he cannot take his garden with him to interest the doctor who fails to take his condition seriously, with disastrous consequences.

Norman's story is based on the few sketchy details I'd learned about my paternal grandparents in the late '50s. Black migrants in England, before and during WWII, had mostly lived in port cities. Working-class whites in the geographical and industrial heart of the country – the West Midlands – were less used to immigrants of colour and felt threatened by the new competition for housing and jobs. Attitudes hardened as the numbers grew. Enoch Powell – a Tory

MP who had worked in Barbados recruiting nurses for the NHS – delivered his 1968 'Rivers of Blood' speech in Birmingham, based on 'concerns' raised by forgotten white people in his Wolverhampton constituency. Another Tory MP, Peter Griffiths, swept the 1964 Smethwick election with the slogan, 'If you want a nigger for a neighbour, vote Labour'. My grandparents' generation toiled to find liveable accommodation, suitable employment and a foothold in a society that didn't want them. The English Dream, instilled by their colonial education, remained perpetually out of reach, but disparities between the England they expected, and the one they received, could at least be resolved in their gardens. They could dig up a little bit of English soil and be responsible for upholding tiny patches of its beauty. They could fit in without drawing unnecessary attention to themselves, retain a sense of the back-home and pride in their heritage. Backyards could be whatever they were required to be – places for children to play safely, for sheets to air, for herbs, strawberries and tomatoes to grow. Front gardens did not speak with a Caribbean accent or appear obviously alien. Passers-by would peer over the wooden fence posts and find England, with the same green grass, butterflies and bees.

As for my maternal grandparents, fourteen years after moving here in 1957, they and their five children were finally awarded a council property that they didn't have to share with other families. Ten years hence, they moved to the house my grandmother kept for thirty-eight years, the council reclaiming it a month after her death. She and my grandfather, who died in 2002, both came from rural St Elizabeth in the south-west of Jamaica. The eldest of eight, my grandmother was a Sunday school teacher who played piano at her local church – facts I didn't learn until after she'd died – but by the '70s, she was working at a lathe for Westfield Stampings in Dudley, her elegant hands and perfect nails besieged by iron swarf. Daffodils sprung from a little bed up against the front wall of her house. She would cut a couple for a vase. Where there once had been a lawn was now shingle; a few neat, well-kept shrubs, English lavender among them, grew around a single central rose bush. Visiting my grandmother on bright afternoons, I often found her headscarved and bent over at the hip – saving her arthritic knees – twisting out new weeds between thumb and forefinger. When I rushed home to see her, having been told she was terminally ill, I realised she'd lost control the moment I saw the metre-high weeds at her gate, keen as sunflowers. When she died, I felt the loss on multiple levels: the loss of her; her cooking; her home; her link to Jamaica. I've never been, and nor have my parents, and so I felt permanently dissociated from my heritage.

In the Black Country, nature can be liminal, confined to back gardens and the spaces between housing and industrial estates, thick with wiry trees, prickly holly, nettles and bracken, discouraging trespassing but inviting illicit activities, strewn – as could be seen through the metal fence posts round the back of my parents' house – with crushed cans of super-strength lager and screwed-up pages from porn magazines. My parents still live in the post-war semi I grew up in, adjacent to council maisonettes and run-down streets occupied, then, by Thatcher's unemployed. Some parts of this region voted to leave the EU by a 70/30 split. Back in the '80s and '90s, the disenfranchised turned either to religion or to the British National Party, for which my borough was a stronghold. The streets were unsafe for a bookish black boy. I was to come straight home from school and wasn't allowed out otherwise, and I thought it was because my mother didn't trust me. Sometimes, on my way home alone – too late to run when they saw me – I would walk into white thug traps and get beaten. They were notorious on the estate, these white thugs, even as children; I'd done nothing to provoke them except exist blackly, and my father, who had probably received similar treatment growing up, could do nothing to help except read me a comforting scripture.

My mother and I didn't get along, so I stayed up in my two-by-three-metre box room reading Bible literature and writing little fantasies in which someone, a man, took pity on me, saw my potential and swooped down to save me. She didn't go out to work, but as we got older my mother delegated domestic jobs to us. While my twin sisters, four years my junior, helped cook simple meals and water the houseplants, I was tasked with vacuuming from top to bottom, scrubbing the bathtub – and gardening.

At their best, before the mass spread of uPVC and IKEA, the houses on my parents' row were quite neat and unfussily styled, with simple wood-framed windows and doors. The front and back gardens are much changed now, mostly dug up and replaced with easier-to-maintain gravel and pots. Before, hanging over a low brick wall, there was a fast-growing privet hedge that had to be cut back regularly with stiff old shears, shielding a small three-by-four-metre garden, entered through a white iron gate. I would mow our lawn, or when the lawnmower stopped working and we couldn't afford to replace it, get down on my knees and trim the grass with the shears. I'd turn over the rich border soil with a hoe, cutting around the edge of the lawn until my mother noticed how much it was shrinking. The oak front door, with its nine little windows and five-panel fanlight, opened into a small hall. Multi-panel interior doors, promoting light about the house, led to the front room, a dining room and kitchen, the latter housed in an extension built by the previous occupants.

We were poor, and surrounded by poor white people. Our whole street was white, apart from the Indians who owned the two shops at either end. The pub on one corner and the working men's club on the other were both no-go areas for us. We were tolerated, as long as we kept quiet, but our house was broken into regularly; our Kingdom Hall meetings were at the same time each week, so the white thugs knew when we'd be out, though each time they came, there was less for them to take. My parents, who had both grown up in the Black Country, apologised for their blackness. I don't remember ever hearing either of them speak patois. They invited white people into our home and cooked them roast dinners, something my grand-parents would never have done. There were none of the Jamaica wall plaques or other signifiers typical of first-generation homes in our house; mass-produced sub-Constable prints of thatched cottage watercolours hung on our stairs.

The backyard was to be kept neat, nothing more. I don't remember my parents ever taking any pride in it, sitting out in the sun or cultivating herbs or flowers. They didn't need their garden to do what Norman's or my grandparents' did. It was better that our front garden did not draw attention to itself, lest we be accused of thinking ourselves better than our neighbours. My parents didn't need to mask the smells of foreign cooking; they eschewed fresh rosemary and thyme. They joined the majority chorus of dissent against the perceived miasmas of Indian shops. We were used to the lingering odours of the closed factories and foundries, and revelled in the laying of fresh tarmac (although my father, a skip driver, came home every night from work and embittered our nostrils with often chemical landfill). We saw ourselves as English, and so when my mother did on the odd occasion cook chicken, rice and peas, it was as an Englishwoman might, with dried herbs and even a lack of seasoning. She was the aspirational OXO mum, and preferred the convenience of packet sauces or granules. Instead of fresh flowers clipped from her own garden and collected in a vase, there was a basket of pound-shop potpourri on the wall unit.

The backyard was a good size for us to play in while washing blew on the line. A path divided the yard into two unequal halves. The more substantial right-hand side was lawned, and a row of conifers and deciduous trees, one with golden-yellow leaves, separated our garden from the neighbour's. A rockery clung to the corner outside the kitchen. On the left-hand side the grass was patchy and the soil hard and thin, and a succulent tree, like a non-flowering magnolia, stood against a wooden fence over which next door's roses spilled. At the bottom was a rotting garage, accessible via a ginnel round the back, and a sycamore we called 'The Big Tree'. My mother would send me out to gather dead leaves, mow the lawn, water the rockery and sweep the patio. I wasn't allowed back inside until every single leaf had been gathered up. It wasn't so bad in the summer, but I dreaded it in the colder months, binning piles of freezing, wet leaves with my bare hands on a Saturday morning while my mother and sisters sat watching kids' TV in front of the fire.

She used the garden as an extension of my box room, so that she could keep an eye on me while shutting me out of her house. I lost myself in daydreams as I brushed up every last crisp bag, chip wrapping and cigarette packet foil that had blown in, sweeping the pathway all the way down to the back gate. It pleased my mother to see a neat garden, though she wouldn't thank me. I'd love to be able to remember my thoughts. I almost certainly never seriously believed in God, or in the vision of the future I was being prepared for despite my outward shows of piety before the congregation. I was yet to read a word written by a black person, a gay person, a black gay person or a woman (Ruth and Esther were written by Samuel and Mordecai respectively). Nor did I want what people outside the organisation seemed to strive for: a steady manufacturing job, a semi-detached council house, a Ford Sierra, a wife and kids. I didn't know what I wanted, but I knew I wouldn't find it in the Black Country. Still, I don't know if I'd have been able to leave of my own volition. It was for the best that I was pushed.

A new Kingdom Hall was built by the labour of the two congregations that would share it, one of them mine. I volunteered for all sorts of jobs, from cleaning to gardening, just so as not to be at home. I pledged sixty hours per month to preaching the ministry, and made myself available to give almost weekly Bible talks, applying scriptures to the needs of the congregation. Now in my late teens, my queerness could no longer be suppressed, despite it being anathema to my religion and apparently incompatible with my race. Witnesses are strongly discouraged from socialising with non-believers, so I sought the company of members whose attitudes were close to the fringes. It was while looking for my own Land of Milk and Honey through these 'bad associations' (1 Corinthians 15:33) that I was able to find my way out of the Garden of Eden: I was disfellowshipped at seventeen, for getting drunk and asking a fellow member to cover for my whereabouts. By twenty-two, I was living in London as a sex worker and dealing with the effects of sexual assault. I found my Sodom and Gomorrah, then read Proust's volume of the same title. The paradisal cover of the *Look!*

brochure had been scrubbed from my mind, replaced with a dark metropolitan realism and new notions of art and literature.

I now live in a fourth-floor London flat overlooking a garden that I have no access to or responsibility over, which backs onto Hampstead Heath, a godsend that gave my partner and me the lockdown privileges of air and exercise. During a rare spring largely free of noise and air pollution, I became more aware of the proliferation and variety of birds the Heath attracts, and downloaded an app, a sort of Shazam for plants, to teach myself more about the local flora. I know next to nothing about technical gardening, though I would love to have one of my own to experiment in because I understand completely what Norman, and my grandparents, appreciated and loved about their gardens: that whatever traumas might come with trying to settle into a hostile environment, they could be in control of their own space, express themselves and assimilate while showing their individuality and nodding back to their heritage – all of which I get through writing.

I accessed the character of Norman by researching Jamaican flora as it would have been at his time, and fell in love with the promise of the island's beauty. A first trip to Jamaica, to eat mango in the shade of a blue jacaranda, will be paradise enough for me.

Putting the Brakes On

NIELLAH ARBOINE

Growing up I was surrounded by plants. Getting lost with my mum in Kew Gardens, falling out of trees in Jamaica, or running down the paths in a pick-your-own, gorging myself on fistfuls of strawberries. But my mum's allotment was probably where I got to know plants best.

The allotment was my fortress, a place where I remember my childhood in technicolour. My mum has had the allotment for as long as I can remember, which is lucky. Now it's near impossible to get your hands on one in London with waiting lists stretching from a few months to decades. Weekends were spent clambering through the knotted roots and branches of the old mulberry tree, watching tadpoles swirl around among the algae in the pond, or carefully picking red-currants and fat green gooseberries.

It was a twenty-minute walk from our home in south London to the allotment, with my older brother in tow and my mum leading the way. In the summer we'd stop off at Iceland and buy a four-pack of strawberry Cornettos, licking and crunching away down the bustling high street and through the quieter back streets until we reached the giant metal gates. It was on those walks that I got to know my mum best. Sometimes on days when it was just the two of us, she'd tell me stories about her childhood, of the mischief she and her siblings and cousins would get up to, and all the games they would play.

Visiting our allotment was like stepping into the film adaptation of *The Secret Garden* I devoured incessantly as a kid. One moment we were in the noise and racket of the city, and the next we'd turned into our own secluded little corner of paradise.

It felt to me like the only place in London where you could happily say hello to people you didn't really know, without receiving a weird look of concern. It was a community: everyone working away, sharing advice, strolling and peering into each other's patches to see whether their leek stems were as thick and flourishing as their neighbour's.

While my mum was turning over soil, plucking up weeds or kneeling on her mat with the shears, my brother and I would chase down the paths throwing bits of sticky cleavers on each other's clothes. Sometimes, visiting the allotment felt a bit like stepping back in time. There was no trace of technology and there were certainly no toilets then, so my mum would have to dig a hole for me. It was good for the plants anyway, she insisted. Really, there's nothing ladylike about gardening.

If we weren't at the allotment, my brother and I were in the children's nature garden down the road from us. There I learned how to make my own rectangular windowsill pots by sawing planks of wood and hammering them into place with nails, probably at far too young an age to be handling tools. We also learned about survival in the great outdoors: how to put up a tent made with canvas and sturdy sticks furnished from tree branches found in the garden's miniature forest. The experience taught me that there's always wildlife and nature to be found in the heart of London, if you look hard enough.

But then adolescence hit me like a truck. And once I was a grumpy teenager, the allotment and the nature garden weren't cool. Like all my peers, I just wanted to fit in, to see my friends and slowly lose brain cells to my phone all day. Bless my mum – asking soon became pleading, as I refused any offer of a trip to our green utopia. It was always too far or too cold, or I had more important things to do.

It wasn't until university when I was thrust into the Welsh countryside that nature came back to me – it's very hard to avoid it in the folds of Cardigan Bay. By my third year, I was taking solo afternoon walks through the forest by campus into the endless fields with grazing sheep and cows, and along the coastal path north from Aberystwyth. I'd sit on a bench facing the sea, eating my packed lunch, surrounded by blue and green.

I sit snugly in the 'wellness generation', millennials and Gen Zs desperately seeking out new ways to look after our mental and physical health. Whether it is through fitness, astrology or growing plants, we're always trying to reconnect with ourselves. Many of us have grown up with anxiety-inducing technology, all of which can feel disrupting and overwhelming.

I'm very impatient. I struggle to queue, and internally huff if someone meanders slowly in front of me on the street. I have a chaotic number of tabs open on my Internet browser, a sad reflection of the multitude of thoughts whirling around in my head. Switching off is a challenge when you're locked into technology and social media that never stops.

Plants have taught me patience in a world that moves at the speed of light. You can't will a plant to grow, or tell a tree how to bend. You can only nurture and care and wait. Gardening truly is the cheapest form of therapy. Caring for my plants has become a way for me to care for myself, not just in the generic millennial 'self-care'-via-a-bubble-bath kind of way, but really forcing me to put the brakes on. I think of the words of Audre Lorde: 'Caring for myself is not self-indulgence, it is self-preservation, and that is an act of political warfare.'

The mental health benefits of gardening are so undeniable it's a true shame that not everyone in our society has the privilege of getting stuck in. Unfortunately, gardening doesn't always feel that accessible. According to the Office for National Statistics, one in eight households doesn't have access to a garden in the UK. Lockdown revealed further inequalities: Black people are four times more likely not to have an outdoor space compared to white people.

Currently, I don't have a garden. Looking for housing in London is already a special type of agony, and it follows that often you have to let go of the dream of outdoor space. Even if you do have access to a garden or a terrace, renting is far from stable. It's a challenge to put down roots when you don't know how long you're going to be somewhere. So it's no surprise that millennials are more likely to have house-plants in every room compared to older generations.

Houseplants manage to make me feel grounded even when the ground isn't my own.

Plants now grace every corner of my flat (bar the kitchen, which has no natural light). I began by dipping my toe in with succulents, the low-maintenance dream, especially for those who get emotionally invested and feel devastated when a plant dies. Next, I moved on to growing food: tomatoes, spring onions, sage, thyme, basil (unsuccessfully), strawberries, Scotch bonnets and courgettes.

During lockdown, my plants became my sanctuary. They were a steady constant in my life when almost everything else felt uncertain. I particularly love growing things I can eat, and staying at home all day gave me time to branch out and dedicate myself to tending to vegetables, which require close attention and intervention. Not expecting much but taking advice from the founder of Grown Club, a gardening group for women of colour, I scraped the seeds out of a sweet pepper and sewed them into some old yoghurt pots. One by one they began to sprout and unfurl. I carefully separated them out, repotting each pepper plant and gifting them to friends and loved ones in my local area. If I wasn't repotting peppers, I was finding a moss pole for my Swiss cheese plant and gently pinning and tying the large round leaves in place.

Having something to care for, something that depends on me, is the most rewarding aspect of gardening. The average age for first-time parents in the UK now sits somewhere between twenty-eight and thirty-three and even that seems still too imminent for me at twenty-seven. Many millennials don't own property, let alone a pet, and although I am certainly in no rush to have kids, I do have a lot of love to give. Being a plant mum allows me to really nurture something.

I guess my generation of gardeners likes to do things a little differently. Social media might spike our anxiety, but there's a special little corner of the internet for plant parents. I turned to Instagram when I noticed a peculiar asparagus-like thing shooting directly out of the centre of one of my succulents on the windowsill. A tad horrified by its resemblance to some flesh-eating species out of *The Day of the Triffids*, I popped it on my story. 'Anyone know what's wrong with my plant?!' I asked frantically, and within minutes I was greeted with multiple replies – 'It's a flower spike! Must be getting a lot of sunlight', 'Totally normal, here's a picture of the full-sized plant' – putting my mind at ease.

There are apps like PictureThis to help you identify plants, ideal for someone like me who doesn't always know my *Haworthia cymbiformis* from my *Haworthia cooperi*. And spending days on Pinterest creating pins for all the houseplants I can't justify buying quite yet has become a hobby in itself.

Although we're separated by distance now as I live a few miles from my family home, I often WhatsApp my mum photos of plants I don't know enough about, and she always responds with the perfect advice. Walking home recently, I found a waxy-leafed houseplant sitting unloved on the roadside. Of course I had to rescue it and bring it home to join my other plant children. My mum didn't know the name of the plant or genus but it was tropical, she concluded. 'Cut off the dead bits,' she said. 'Talk to it.'

Last spring, I went back to my mum's allotment for the first time in eight years. Like many people across the world, flung into chaos, I hadn't seen my family in months because of the pandemic, and the cold chill of isolation had really kicked in. In line with the lockdown rules, somewhere outside was the only place I could see my mum. Of course, we chose the allotment.

Everything was almost exactly as I'd left it years ago. Well, with a few new additions: a toilet, for one, and a shared cabin for announcements and flyers. Returning as an adult felt odd – nostalgia punctured by reality. The mulberry tree I'd clambered up as a child seemed smaller, older. But as we turned onto our plot, the colours of spring adorned every corner of our patch. Plump golden bumblebees flew through the air, bulbs looked ready to burst open with life, and the shades of green created a luscious patchwork across the ground. I hadn't noticed it until then but over time, my mum's fingers had only become greener. To see what she had achieved on a modest-sized plot was truly astounding.

The allotment was more magical than I recalled. Everything was alive and unpredictable. Mum had spring onions, potatoes and green beans, which, she said, were being 'ravaged by the wood pigeons'. Something had gone wrong with the old gooseberry bush, but she had another one growing now. She was also growing sorrel: 'But I'm leaving it to go to seed.' There was lemon balm, chive and black sage. Sadly the rhubarb had gone all funny, but she didn't seem to mind. That's just the nature of plants sometimes.

'I can't remember if it's twelve or eight, but if you have a particular number of fruit trees it becomes an orchard,' my mum told me, as we sat on the bench at the back of the allotment sipping champagne and eating pasta out of Tupperware perched on our laps. The afternoon sun touched everything with gold. 'But it might be only apple or pears,' she pondered. We counted the miniature trees in front of us: two cherries, apples, pears, plums and some more apples. It looked like at least ten trees to me.

'I'm starting to use nettles now,' she continued between sips, describing how she keeps the nettles she would usually cut back and compost, mashing them in a bucket. 'I'm reading about a lot of things that we consider weeds.' She went through the ingredients of her potassium- and nitrate-rich 'fertiliser tea', as she calls it, a brown swampy concoction which she pours over the plants to help them grow. 'It's disgusting to look at, but it's all about using everything you have – no wasting.'

My mum's always shown me how to be sustainable, sort of through osmosis. Never throw away the potted herbs from the supermarket; make compost from your leftover peels and scraps; stick spring onion roots in water so they sprout; and if you've grown a surplus of fruit or veg, pop it in the freezer (in a recycled ice cream container, of course).

I still have so much to learn about gardening, and a lifetime of lessons to be taught from plants. Everything sits at the end of our fingertips through our phones and technology. But this past year, more than any other, gardening has shown me that instant satisfaction isn't always what I need. It's time for me to practise patience.

The Language of the Garden

Looking at the Garden

CLAIRE LOWDON

My mother is a gardener. We had a big garden when I was growing up and for much of my adolescence we were close to self-sufficient in fruit and vegetables. Onions, garlic, leeks, potatoes, peas, beans. Espaliered apple trees bearing varieties you never saw at the supermarket. I liked the Egremont Russet, with its skin like fine-grade sandpaper and that faint flavour of peach. Asparagus, and too many artichokes to eat: the ones that got away flamed ultraviolet. At the centre of it all there was a hop, totem of my mother's own childhood on a farm in Kent.

The potato crop was kept in old filing cabinets with twists of newspaper to stop them from touching. Carrots were stored in sand. Fierce little chilli peppers were pickled, then eaten straight from the jar with a cooling schmear of Philadelphia cheese. For the second half of summer, the freezer was precarious with baking trays of blanched French beans. My three siblings and I could supplement our pocket money by squashing caterpillars: 1p for a highlighter-green grub that oozed olive-coloured gunk, 5p for bringing a thumb down on a pixelated cluster of eggs. We ate nectarines blood-warm from the greenhouse, the same fruit we'd helped pollinate with paintbrushes earlier in the year.

My mother in her night dress in the morning, chasing the rabbits away.

Or murderous with a torch after dark, the stinking slug bucket a-slosh at her side.

The magic trick of a summer supper in December: basil oil in your tomato soup, slim young carrots sweet as candy, strawberry granita for pudding.

There were flowers, too, and in the evenings after school my mother would walk me round the garden and point out the new arrivals. The first of the oriental poppies, petals wrinkled from the bud, salmon-pink paint still drying. The complex cryptogram of a pencil iris that wasn't there yesterday. Extremely local miracles, but my mother greeted each one like a real happening.

I went along willingly enough, although I wasn't that interested in the plants, at first; the draw was my time alone with her, the privileged entry into an adult domain. Gradually the garden came into focus. I don't remember her actively trying to teach me anything, but some of it got learned anyway. So many of the names made such beautiful sense. Lungwort, or pulmonaria, has lung-shaped leaves; *pulmones* is Latin for 'lungs'. Foxglove is *Digitalis*, because its flowers would fit so easily over your fingers, or digits. (Checking it now, I learn more: the late eighteenth-century Latin name was inspired by the older German *fingerhut*, which means both 'thimble' and 'foxglove'.)

At school I was a fairweather student of Latin, bored by the grammar, greedy for the vocab. Making the connection between 'amo' and 'amiable', or linking up pescatarian-pisces-pêcheur-fish, gave me a deep, atavistic thrill. With plants it was even better, because encountering the language also called the visible world into being. In 'How the Alphabet Was Made', my favourite of Kipling's *Just So Stories*, Taffy and her Daddy Tegumai invent letters by making sounds and drawing their mouth-shapes on pieces of birch bark: 'ah' becomes the gaping carp fish 'A'; 'O' is round like an egg or a pebble. 'I believe we've found out the big secret of the world,' Tegumai says. The garden was full of secrets, and names were the keys that unlocked them.

As I gained a basic grasp of the plants on our evening rounds, I found myself naming them aloud too. Not to show off my new knowledge (though that wouldn't have been atypical), but for the pleasure of saying *I see*. No different, really, from a much younger child calling out when she spies a dog. Before, there were just stalks and leaves and flowers. Sure, some were bigger than others, there were pink ones and blue ones and orange ones. But when I learned a plant's name, it became truly distinct. You couldn't mistake Jacob's Ladder for aquilegia once you knew to look for the stepwise leaves. (And aquilegia, I now discover, takes its Latin name from *aquila*, 'eagle', because of the claw-like aspect of its spiky flowers.)

Most of us have plant blindness to some degree and each generation is more myopic than the last. This is problematic for descriptions of plants in literature. Reviewing John Updike's early novels in the *TLS*, I wanted to showcase his talent for nature writing. Quoting from *Rabbit, Run*, I thought I could just about get away with referencing 'the shaggy golden suds of bloomy forsythia [glowing] through the smoke that fogs the garden'. But reluctantly I set aside this exquisite miniature of lily of the valley flowers: 'the high ones on the stem still the faint sherbet green of cantaloupe rind'. Nor did I include Rabbit's riff on rhododendrons: 'when the hemispheres of blossom appear in crowds they remind him of nothing so much as the hats worn by cheap girls to church on Easter [. . .] each individual flower wears on the roof of its mouth two fans of freckles where the antlers tap'. I wasn't certain that enough readers – especially younger ones – would be able to visualise lily of the valley and rhododendrons. The gratification of the imagery depends on recognition.

Yet even if you don't know what they mean, straight lists of plant names can be oddly evocative. In Tracey Chevalier's *At the Edge of the Orchard*, set in mid-nineteenth-century America, the flora of an Ohio swamp is swiftly conjured: 'he sat and looked at the green fuzz taking over the trees, at the birds flitting through the branches as they built nests, at the trout lilies and trilliums and Dutchman's breeches at his feet'. I have a rough idea of what a trillium might look like and no clue about the other two; but the names provide enough specificity for that not to matter much. And how many of us could confidently identify more than the nettle in King Lear's mad headgear?

> Crown'd with rank fumiter and furrow-weeds
> With hardocks, hemlock, nettles, cuckoo-flowers,
> Darnel, and all the idle weeds that grow
> In our sustaining corn.

The magic of it is that when I stop to look these plants up online, I realise I've seen most of them before. I've just never *seen* them. Now that we've forgotten so many, botanical names can function like reverse-engineered clichés, revenant bits of language that teach us how to look. 'Fumiter', I learn, is from the Latin for 'smoke of the earth': hazy grey-blue foliage, slightly translucent flowers, rising from the ground in great profusion. Newly heard, the name works in the same way as good writing: it sharpens my perception of overfamiliar reality, just as Updike noticing that precise melon rind colouring augments my appreciation of lily of the valley.

All those after-school evenings with my mother were formative lessons in language as well as botany. Most of all, they were lessons in looking: early proof that there is always more to see. I remember exclaiming, in my effusive, sixteen-year-old way, about a tiny vase of snowdrops on the kitchen table, the delicate green pinstripe on the underside of each petal. And my mother – who is by nature much less effusive, verbal, demonstrative than I am – smiling tolerantly, and then saying with a strange sideways look: 'If you can feel that way about a snowdrop, you're going to have a happy life.' It was an uncharacteristically personal pronouncement that I didn't fully understand at the time. I grasped just enough of what she meant to receive it as a secret conferred, a blessing.

Sensibly, heartbreakingly, my parents sold the house when I was in my early twenties. Only my youngest sister was still at home and soon she'd be gone, too;

the place was too big and old for just two people. We all understood but it felt like being evicted from Eden.

I'm lucky enough to have my own garden now, a long thin strip at the back of a terraced house in Oxford. When I moved here with my husband in late 2016 we had no furniture, not even a bed, but I had sweet peas germinating in the boiler cupboard by the end of week one. After years of titivating succulents on other people's windowsills, I couldn't wait to get started. The garden itself was a permissive blank: just a scrubby lawn and too many overlarge trees. Down came the trees, up came the lawn, or most of it, to make way for flower beds which in the first year I stuffed excitedly with annuals. (An early beloved gardening book was James Fenton's pocket-sized gem, *A Garden from a Hundred Packets of Seed*.) Honeywort, clary, borage (a mistake; it's a virulent self-seeder that I'm still pulling out four years later). Nasturtiums, violas, cosmos, sunflowers. Lots of marigolds: the simple saffron-coloured 'Indian Prince'; showy 'Jolly Jester' with its burgundy and yellow stripes; and my favourite of all, the Swedish marigold Burning Embers (or *Tagetes Linnaeus*, after Carl Linnaeus, the father of modern taxonomy), its orange velvet petals singed with gold.

That first year I was all theory and no practice: I'd often helped my mother in the garden as a child, but

I'd never nurtured a plant on a daily basis all the way through its life cycle. It is a miraculous, mind-blowing process, and growing annuals is the fastest, most startling way to experience it. You begin with a seed the size of an eyelash and in a few months you have a three-foot-high, bushy plant that can flower from June to November. Mostly the plant does this itself. OK, so you have to water it, pot it on a few times, plant it out in the ground. But it's more or less unstoppable. Just compare that growth rate to a mammal's.

I'd deliberately chosen 'easy' seeds and almost all of them 'worked'. I was delighted, euphoric, likely rather boring on the subject to friends and family. But I was also a little freaked out. It all seemed so ... unnatural. Of course I do know that very little of the nature we see around us is 'natural', because humans have been intervening in the sex lives of plants since forever. The garden annuals we buy as seed have been selectively bred for fast replication. Still, it's very easy to forget – or to never realise – the sheer power of the natural world, as we stroll along our pavements, shaded by the boughs of conveniently planted sycamores. Those roots pushing up through the concrete: they should be a clue, but mostly we just step over them and move on. Not me anymore, at least not in the summer of 2017. Now that I had witnessed first-hand the awful might of a marigold, I had a better idea of what a tree might be capable of.

The other newly creepy thing about trees: how similar they were to many of the plants in my garden. Same central stalk, same branching aspect, same surprisingly tiny root ball. My childhood introduction to gardening had been a game of spot-the-difference. But once you know that rosemary, sage and mint are all members of the salvia genus, the family resemblance starts to become obvious, even sinister, those differences mere variations on a theme. And the rosemary in my herb bed looks just like the rosemary next door. Returning to the garden in my thirties, it was this sameness of plants that I saw afresh. Afresh: a word I can never write or say without hearing the drum beat of Larkin's clear-eyed spring poem, 'The Trees'. The speaker of the poem fingers the lure of spring: the trees coming into leaf seem to herald the death of the old year, tempting us to 'begin afresh, afresh, afresh'. Yet the trees are like us, the speaker insists; they are also getting older, approaching death. The mounting number of rings in their trunks belie their apparent rebirth. I had always assented to this, but now I wanted to object. Plants don't really die, they all look so alike they might as well be cloning themselves! (And some of them actually are: blackberries, dandelions, crab apples and many others reproduce asexually.)

In 1484 the Flemish artist Hans Memling painted a triptych commissioned by Willem and Barbara Moreel.

The Moreels are depicted in postures of devotion on the two outer panels, facing inwards towards the central panel where St Christopher holds the Christ child aloft. No one is looking at the ground, but Memling clearly was: he has given us, in brilliant detail, dandelion and plantain, wild strawberries and daisies and celandine. While Willem and Barbara with their matching cleft chins have been dead for over five hundred years, the plants persist. You can see them all – apart from the wild strawberries, perhaps – on any country walk in April. How extraordinary that something as fragile as a daisy should be so much more durable than we are!

This is remembered wonder, remembered awe: three years on I've got used to it all over again. Now I tut impatiently if my annual seedlings are slow to bulk out, forgetting how insanely fast the whole process seemed to me only recently. Instead, I notice other things. The garden is filling up with perennials, some carefully nurtured from seed, others purchased, thrillingly, by post. Watching their various growth cycles come round again each year is making me properly aware of the seasons: it's no exaggeration to say that the way I think about time has changed. Before, I could easily get distracted and find myself in the middle of autumn without realising it: my goodness, is it five o'clock already? Inspecting the garden each day is like watching the second hand go round, and it's glorious.

That early lesson: there is always more to see. Nabokov put it best. Musing on the subjective nature of reality in an interview in 1962, he noticed the different ways people perceive the same natural object – in his example, a lily. A naturalist experiences a lily more intensely than an ordinary person does. Someone specialising further – a botanist, or even an expert in lilies – will experience the same flower in sharper focus still. 'You can know more and more about one thing but you can never know everything about one thing; it's hopeless.' I say hurrah; I'm all for shocks and false bottoms, for seeing and re-seeing over and over again.

In the first weeks of lockdown, so many people reported noticing nature more. More beauty, more birdsong. I think I mostly noticed nature itself *not* noticing: just getting on with the business of burgeoning, naturally not giving a fuck that one of its myriad species is sick. (Think of the torturer's horse in Auden's poem 'Musée des Beaux Arts', scratching its 'innocent behind on a tree'.) Our garden has been doing its usual thing too, the tulips giving way to the alliums and irises, the scabious and the clematis and the foxgloves. The big excitement this spring has been a rambling rose we planted as bareroot stock; it's just flowered for the first time after three years of gathering strength. And what strength! It must have put on four feet in two months, not to mention the masses of flowers, creamy double blooms with a tassel of gold braid at their centres.

I've tried taking my mother on a few virtual tours during lockdown, but FaceTime is a shaky, unsatisfying way to experience flowers. Photographs shared on WhatsApp will have to do for now.

Me: Expert advice needed. This is a massive weed, right?
Her: Massive weed. Make it history.

In May, my Russian sage seeds popped up in just three days, despite the packet's stern warning: 'Germination is slow and erratic taking 3–12 weeks approx.' Jubilant, I sent a photo, which earned a phone call in return and a moment of perfectly shared excitement: utterly uncomplicated mutual wonder at the natural world. And also: 'Do you remember I used to grow Russian sage in the garden?' In fact I'd forgotten. Before buying the seeds I'd looked up the plant online because I liked the name; the images were familiar but I couldn't quite place it. Her memory made me look again at my own memory and see the plant there after all, its unearthly blue spires only half visible in the gathering dusk. 'Yes,' I told her, 'I remember.'

Five Tongues

DAISY LAFARGE

First Tongue

The first tongue is a legend, a hearsay, a heresy.

We moved to the house with the garden of five tongues in early April. The garden was my first and had, for some years, been left to its own devices. These devices seemed to favour – almost exclusively – a plant I did not recognise, but which grew wherever I looked: in full sun, teetering over the paving stones, or clumped beneath the shade of the eucalyptus tree, which shed its reddening leaves indiscriminately. These plants had tall, confident stems with fractal green leaves, small lilac buds that bloomed into blue. The garden had its devices and so did I; among them, an English affliction of nouns, intensified by enforced domesticity. These plants would be my contemporaries for the foreseeable – I wanted us to become acquainted.

But I didn't know the plant; my new neighbours didn't know the plant. At first even the internet didn't seem to know the plant. Forums kept throwing up chimeric approximations of a pervasive weed with forget-me-not flowers, comfrey leaves and a nettle-like sting. A few posts mentioned its attractiveness to bees, but on the whole the plant was considered a pest, a weed. The plant's pernicious roots were mentioned regularly, almost effusively. You have to dig out the whole taproot, the posts warned, and even then you'll be lucky to stop the plant reseeding next year.

Some were calling it alkanet, but 'alkanet' also seemed to refer to a variety of plants, many of them in the *Boraginaceae* family, particularly the dye plant, *Alkanna tinctoria*, which the plants in my garden were not. Nor was it another alkanet, common bugloss or *Anchusa officinalis*, although the flowers bore some blue, familial resemblance. Perhaps it was this other word, bugloss, that led me to the name – names, nouns – of my plants. Searches for 'evergreen bugloss', 'alkanet' and 'green alkanet' all turned up the small bright flowers I could see through the window: *Pentaglottis sempervirens*.

Pentaglottis sempervirens: its five tongues are Greek (*penta* + *glottis*) and its evergreen, 'always-living' is Latin (*semper* + *virens*). Its leaves are borage blue but its common name is stained red with henna; alkanet derives from the Arabic *al-hinnā*, muddled through Medieval Latin, Old Spanish and Middle English.

No one knows exactly how or when it got here (the plant, not the name). It is native to south-west Europe and was allegedly introduced to the UK before 1700, perhaps mistaken for its relative, dyer's alkanet, *Alkanna tinctoria*, or else as a cheaper alternative to its other erroneous namesake, the true henna plant, *Lawsonia inermis*. The first recorded sighting of green alkanet in the 'wild' dates to 1724, but it now seems to be endemic all over the UK; its dislike of acidity makes it particularly suited to the clay-based soil of cities.

Online I buy a botanical print of green alkanet, a page that has been carefully excised from a book dating to 1837. Along with the illustration (confusingly labelled *Anchusa sempervirens*) I am sent another page of the book concerning the plant's taxonomy, general features, locality and listed sightings: 'In waste ground, among ruins, and by way-sides' (Miss Armetriding); 'Emmanuel College, in the Master's close, under one of the walnut trees' (L.W. Dillwyn Esq.); 'Road-side at Walthamstow, Essex, possibly from a garden' (Mr E. Foster); 'Deanbank, near the village of the Water of Leith' (Dr Parsons); 'Banks of the Clyde, above Hamilton: near Chatelherault and Castlemilk, Glasgow' (Mr Maughan).

I follow a scant trail through the intervening centuries: In 19th century Wiltshire agricultural dialect, alkanet becomes 'isnet'; Kew Gardens' bulletin of 1975 describes it as being in flower outdoors in January; on June 17th 1990, Derek Jarman counts alkanet among the flowers he sees on his walk, and again on July 4th when he walks along the shingle beach of Dungeness at sundown.

Part of the difficulty in tracing the plant's history is its elusion and illusion of names; it has even managed that rare trick of shrugging off the name Linnaeus gave it: *Anchusa sempervirens*. It was later christened *Pentaglottis sempervirens* by the botanists Tausch and L.H. Bailey, but this belated clarification seems only to have further confused matters; the blogs and forums I read online are oddly unified by calls for botanists to shed light on the 'true' name of this plant that both is and isn't henna, alkanet, anchusa, bugloss, and borage.

This elusiveness seems even to precede binomial classification: Theophrastus' *Historia Plantarum* (4th century BCE) mentions one general variety of anchusa, while Dioscorides lists three kinds. Most classical sources seem to refer to *Alkanna tinctoria*, dyer's alkanet. In the 16th century, the renaissance physician Franciscus Frigimelica distinguished anchusa or true alkanet from the similar bugloss or borage. He notes that bugloss is frequently used to warm the heart, and that the roots are good for melancholy.

Other members of the alkanet family, such as comfrey, are also believed to have medicinal importance. In the classical principle known as the doctrine of signatures, you can discern a plant's healing properties from its resemblance to body parts; *Pulmonaria* or lungwort is another relative of green alkanet, the pink-purple tubular flowers of which are thought to resemble human lungs, and contain antibacterial properties that protect against chest infections and other pulmonary ailments.

And what of green alkanet? I go outside and crouch down next to a plant, trying to see human organs or body parts in its morphology. Its five tongues – which I presume are the five-petalled flower heads – don't seem to correlate to any part of me. In my eagerness to identify with the plant I reach out and touch a leaf, forgetting about the sting, which thuds into nerve endings on the back of my hand. Later I read that the stings of nettles and green alkanet alike are caused by cystoliths, molecular concentrations of silicon dioxide and calcium carbonate. The sting on the back of my hand is lithic, the scratches of botanical stones.

Fourth Tongue

The fourth tongue is a rhythm, a reciprocity.

Still raining, and the last patch of green alkanet wilting beneath the eucalyptus, aphid-heavy. Before the rains started, and before the aphids had tipped over whatever equilibrium existed between them and the alkanet, I had unexpectedly interrupted a family gathering.

It was late April. I picked a blue flowerhead and placed it on a slide. Through the microscope I found myself eye to eye with an enormous mother aphid, her body fattened (I remember reading that black ants farm aphids and milk them like cows) and the colour of pale celadon. She was motionless, poised near the centre of the flower that crystallised an intense, copper sulphate blue around her. Everything was still until I noticed another aphid, a fifth of her size, clambering up towards her. This aphid – who I presumed was her young – seemed to embody the neotenous features of cartoonish cuteness: huge, wide-set eyes, and spindly, curious antennae that probed about endearingly as it approached her.

As I watched them, words like 'tender' drifted across my mind, refusing to dissipate. I was bringing other words too, like 'mother' 'family' 'gathering' and 'child'. I knew this had more to do with me than the aphids, and that in another mood I might have brought words like 'prey' 'victim' 'attacker' 'cruelty'. I knew that words, in general, weren't advisable, but I had never known how to stop. I wondered if the way I should be looking – with scientific impartiality – was any less partial or rife with projection.

Regardless of the scene's true essence, I soon felt like an invasive species myself, and put everyone – flower-head, aphid and aphid – back outside.

Later I read that six varieties of aphid have been recorded on *Pentaglottis sempervirens* in the UK, and one of them, *Ovatomyzus boraginacearum*, is particular to the borage family of plants. From the grainy footage I had taken of the mother (sorry) aphid, *O. boraginacearum* seemed to best resemble those currently overrunning my plants. When I searched for 'aphid and alkanet symbiosis' what turned up instead was the aphid's symbiosis with a bacteria that lives in its hindgut.

Buchnera aphidicola, I read, has lived in a symbiotic relationship with aphids for 160-280 million years. Prior to this, it is believed that its ancestors lived freely in the environment, similar to *E. coli*, with which it shares a seventh of its genome. In the intervening hundreds of millions of years, aphids and *B. aphidicola* have become mutually dependent: the bacteria provide aphids with essential amino acids missing from their diet of plant sap, while the aphid's hindgut provides the bacteria with a warm, stable environment, outside of which they can no longer survive. As a result of this long-standing relationship, *B. aphidicola* has lost a large swathe of its genes, leaving the bacterium with one of the smallest (but also most stable) genomes of any living organism.

I look back at the plants and wonder if their relationship with the aphids is heading in a similar direction. How long does it take for two species to be so symbiotically enmeshed that one of them begins shedding genetic material like old skin? I find myself worrying – on behalf of the organism with a shrinking genome – about misplaced trust in this dynamic. What if the host you believe will support you indefinitely, begins to behave otherwise? By which point you have already jettisoned the adaptive faculties that would enable you to live without it?

In another paper I read that scientists believe this particular mutualism between aphids and bacteria may aid in greater understanding of symbiogenesis. Symbiogenesis, popularised by Lynn Margulis, posits evolution as a frenzy of illicit mergers: billions of years ago, free-living cells were engulfed by other cells and began to act like their organs, a series of integrations which led to the development of multicellular life. If the traditional model of evolution is a tree, symbiogenesis reveals the thick meshwork abstracting its shape, where roots rise up and latch onto branches, and flowers grow along dark mycelial paths, blooming underground.

With symbiogenesis in mind, I wonder whether *B. aphidicola* has become – or is on its way to becoming – one of the aphid's organs. And if our own lives result from this history of multispecies mergers, whose genetic residue we carry in our bodies, then the concept of distinct and separate species is troubled: we are all already interwoven, twined in various stages of becoming each other's organs. The unsuitability of the tree model is perhaps also the failing of taxonomy: the notion that one branch will grow obediently from another. But the branches are untameable and wayward, swapping names and genetic material while the botanists' backs are turned.

Fifth Tongue

The fifth tongue is a signature, a stitch, a lover.

It was July, and still raining. The sea of alkanet had receded from the garden, except where new young shoots began to reappear in the beds we had cleared for other plants: nasturtium, love-in-a-mist, marigold, red velvet sunflowers. Many of these seeds didn't take, and when we inspected the new green growth in the beds were resigned to find, once again, the downy, acerose leaves of alkanet. It seemed – as warned – that we had been unsuccessful at removing all the taproot; the plant would be with us for good.

I was beginning to accept the intractability of what I had tried and failed to retrieve from the alkanet: colour, a fixed name or history, even its rooted presence in my garden. I bristled at these methods, echoes of the colonial impulse to order, extract and exploit. I was intrigued by something I read in a book at this time: '. . . the botanists [. . .] emptied worlds of their names; they emptied the worlds of things animal, vegetable and mineral of their names and replaced these names with names pleasing to them'.[1]

Naming, I had thought, was a kind of filling, a closing up of gaps. But alkanet had made me think otherwise; taxonomy was not a filling but an emptying, a draining out. I wondered if etymology – or at least, demanding a logical trajectory from it – was the great attachment disorder: a longing to close the gaps and fill in the blanks, but in doing so, only driving more space between its parts. Maybe this aligns to what Lyn Hejinian calls the 'rejection of closure', that our failure to close the gaps in language, the loops between what we mean and what we say, *is* the experience of desire; how we dwell in its striving. Like language, the perfect garden is always incomplete, in a state of longing to become what it isn't yet, but could be.

I read that, in his forming of binomial classification, Linnaeus strove too; he dreamed of something more gestural and abstract, a language of botanical calligrams, in which the shapes, properties and behaviour of individual species would be mirrored and denoted by their corresponding symbol. Planets, elements and music all have their own form of notation, so why not plants? I began to dream, after Linnaeus, of a botanical solfège: not the staid symbolism of Victorian floriography, but an anarchic biosemiotics in which signs are fluid and various, mingling in their contradictions.

At some point the lines of blue pen in my notebook – belonging both to words about alkanet and sketches of alkanet – began to form their own illicit, symbiotic merger. The blue ink was looping somewhere between a word and a drawing, it was writing but also not writing, it was an ampersand (&), a symbol that had at one time been the 27th letter of the alphabet, born of a merger of 'e' and 't'; it was later removed from the alphabet but the ampersand had already become its own organism.

This was the organism I saw all over my garden, and at roadsides, and the edges of pavements:

&&&&&

A calligram of loops; sutures; roots and arabesques; leaves; flowers; infinity with a tail or baggage; a figure of eight spilling its guts; a sinuous commons; sensuous typography; spicules that sting to touch.

Alkanet was ampersand; the more I looked at the former the more it began to encompass everything I wanted to say, or couldn't say, about the latter. If a garden full of names – closed gaps – is empty, a garden without names is brimming with kenotic potential. I had attempted five tongues of knowing and arrived at this symbol. At worst it was just another name; at best it was a species of lover.*

* 'A metaphor is a species of symbol. So is a lover.' – Anne Carson

A Ghost Story

ZING TSJENG

I must have watched my mother disappear a thousand times. Every few years, she would sink into the inky murk of her depression, and then kick her way up to the surface. Electroconvulsive therapy helped, but it blitzed holes in her memory. Decaf coffee and Xanax let her sleep. Her psychiatrist became an old family friend, whom I also saw for my teenage depression, a short-lived interruption compared to my mother's episodes, which built up slowly and imperceptibly, like shifting tectonic plates that periodically erupted. Years later, a university counsellor told me this was an 'unhealthy' psychiatric arrangement. Furious, I never saw the counsellor again.

I saw my mother go under so many times, staring

at the TV, cooling cup of coffee on the floor, another taped rerun of *The X-Files* – so it came as a shock when she almost actually did die, in the physical sense of things. Years of chain-smoking, taken up when she was a schoolgirl, paid off in her sixties and she went into hospital on the verge of a heart attack. The 'on the verge' part was important, her doctor told me, as she probably would have died if she'd actually had a heart attack.

As it was, they discovered her clogged-up heart just in time to wheel her into surgery, where she almost died on the operating table. This was the woman who'd told me repeatedly that she would rather buy a one-way ticket to Dignitas than eke out a slow physical and cognitive decline in old age, so maybe this was more of a trial run to Switzerland than she let on.

I flew out to Singapore over Christmas and saw her in ICU. Her eyes were closed and her skin was the colour of ash trodden on concrete. 'You know, we all smoked at fifteen,' her friend told me once, 'but your mother was the only one who kept it up.' When she was discharged from ICU and admitted into her own hospital room, I stayed with her overnight, but she kept waking up in pain, trying to pull out her IV drip, confused and shouting. The ward was horror-movie dark. I was terrified; I tried to go over and comfort her, and she looked directly at me and called me by my brother's name, which hurt in a very sharp and specific place in my chest. Eventually she fell asleep; I imagine the whole night disintegrated over the next few hours in her memory, already frazzled by bouts of ECT.

Anyway: gardening.

According to my mother, if your plant needs flowering, you should steep banana peel in water over the course of a week and then dilute this liquid one part to five parts water, and then water your plants with the mixture. This always works for her orchids. If you have a barbecue, you should keep the charcoal ash and then mix it with soil around the base of your plants to fertilise them. And my mother's motto for gardening? *What's the worst that can happen? It'll grow back.*

I never cared about gardening until my mum's almost-heart attack. I had pot plants, but they were meek, docile things; funky-shaped succulents and tiny cacti, plants that needed no love and thrived on neglect. When I moved into a house in London with a garden, my mum was the one who came over on holiday and pruned it all back, revealing a small Japanese maple tree, rambling roses with teacup-sized yellow flowers and a malnourished olive tree.

The Japanese maple was her favourite – the Acer tree came to about her chest and had drooping, red leaves that looked like they were weeping rusty blood. She tended it religiously, clipping off the spindlier branches and packing it with fertiliser. Objects appeared in my garden – a tool box, a bubblegum-pink gardening stool, a pop-up bin for foliage. Then birds: tiny wrens, sparrows, occasionally blue tits – and animals: squirrels, the neighbourhood ginger cat, purring on the top of our fence and begging us to let her in. Bags of compost were hauled into the garden – despite being in her sixties, my mother is still the strongest person I know – and once I came home to find my six-foot-high monstera plant repotted into something big enough for me to sit in, its roots gently exhaling after its release from the plastic enclosure I'd bought it in.

She took me to a garden centre – all this time I'd lived in London, and I didn't even know where my nearest garden centre was – and we pushed an enormous trolley and stacked it with plants I didn't know the names of. There was an entire courtyard dedicated to herbs, where the scent of apple mint and lavender perfumed the air. Sometimes we walked through the gentle mist from a garden centre employee hosing down the plants. The air smelled of loam and earth, that after-the-storm dampness that makes everything smell so fresh, like God had just turned you out of the Garden of Eden. It reminded me of Singapore, where the humidity builds in a sweaty, sour cloak and finally tears out of the sky to wash everything away, all rain-kissed and new.

Selfishly, I let her get on with my garden on her own. *A nice holiday activity*, I told myself. *A retirement hobby for old people.* Then she left for Singapore and her almost-heart attack followed, and so did I, flying back for her operation. When I returned to London, I found myself crying in front of that Japanese maple. I'd left it alone too long during my Christmas in Singapore and that night in the hospital; hadn't given it the right amount of water, or coddled it with fertiliser, or simply hadn't talked to it enough, in that low tender voice my mum always used with plants, half talking to herself and half talking to whatever was in her gloved hands, so raw and green. Either way, it was dying. Its leaves were almost burnt away to a pale crisp, drooping in an unnatural way. Half the branches were bare.

Obviously, I knew exactly what was happening here – after all, I'd had therapy; even if I was in an 'unhealthy' psychiatric arrangement, I could tell I was projecting. A mother isn't a tree, and a tree isn't a mother. It wasn't even the right colour or shape to be my mother, who was sturdy and brown in a way that an Acer isn't. But the brain works in clichés – you can try to fight it, but no matter how much European cinema you watch, it'll always be the Hollywood romcom that makes you cry.

My mother resurfaced after a week on the hospital ward, fighting her way back through the fog of general anaesthesia and requesting griddled kaya toast and dark, sweet coffee. I told her about her mistaking me for my brother; she said 'oh dear', and laughed. I surprised myself by laughing too. The Japanese maple recovered, and my mother came back to my garden in London to plant it in the earth, where it took root and continues to grow. A happy ending, I guess. But that dark night in the hospital sometimes replays in my mind; the way she cried out in the night and rose out of the bed like a reanimated corpse.

When I was little, my atheist mother – in one of her acts of forgetfulness – signed me up to attend a highly religious primary school which preached, among other things, that the Vengaboys and Harry Potter were works of the devil. Each week, one hour of the timetable was given over to an extracurricular class called 'Christian Values', where the school pastor would sweep in to extol the satanic properties of Tinky Winky and the Pokémon Game Boy.

The pastor impressed upon us the importance of the resurrection and the eventual apocalypse – these were, to her, inextricably tied together. 'Just think,' she said, 'of Jesus waking up in that cold dark cave on Easter Sunday.' I imagined the sudden confusion, the thirst, the coolness of the rocky floor against his skin, the staggering towards the light. And then, a certain amount of time later – the pastor was never clear on dates here – Jesus would reappear for the Second Coming, heralding the end of the world. In my ten-year-old head, the apocalypse clarified itself as an unusually vivid image: my family, their heads aflame like human matchsticks, screaming, roaming the earth like wraiths, eventually ending up in the fiery depths of Hell.

My mother frowned when I told her about this vision and begged her to let us convert. 'What about Gandhi?' she said. *What about Gandhi*, I asked. 'Didn't Gandhi do good things?' *Well, yeah*. 'But he wasn't a Christian.' *Yeah . . .* 'So does that mean Gandhi is in Hell right now?' *I guess.* 'That doesn't seem fair to me. Does that seem like something God would do?' As deradicalisation programmes go, it was brief but effective.

I don't think you ever grow out of those moments of religiosity, though. I see it in people I meet – it's always the ones who don't get the concept of moderation, who ride life hard with one eye on death (that old friend). If you grow up religious, part of you is always concerned with the exit plan. If you spend your life doing good works, it makes sense that you start wishing for eternal rest. It's the same part of your brain that makes you think a tree is a mother, or a mother is a tree; the blurring of physical and symbolic that transfigures water into wine, wafer into flesh, a mother into a corpse and a corpse into a mother.

Growing up, I watched a Chinese period drama on TV where a dutiful son, out of filial piety, followed his dying mother to Hell. He knew that if he released the sleeve of her robe, she would tumble into the realm of the dead and meet Meng Po, the goddess of forgetfulness, who makes the departed drink a soup to forget their earthly life. As long as he held on to her, she would never forget him. So, as a child, I fell asleep pinching my mother's batik pyjama between my chubby fingers. Sometimes I think I am still holding her sleeve.

After my mother and the Japanese maple almost died, I started gardening in earnest. Of course I did. I find the obviousness of the metaphor almost embarrassing; how easily my mind leapt from her getting better to me wanting to plant things and encourage their small, alive possibilities to take root. If my mother ever realised this or felt embarrassed too, she didn't let on.

She just messages me gardening tips, updating me on the status of the mango tree ('no mangoes') and giving out recipes for tea brewed from home-grown lemongrass.

I started growing vegetables: squat orange Nantes carrots, leafy Italian trombocino squash, Chinese kai lan, ragged tomato tumbler plants, vigorous pea shoots and rocket (which always bolted too early in the sun). I grew from seedling and seed; I used my bare hands to sift through dirt and soil until I got a skin infection that manifested as four irregular-shaped black dots on my forearm. It made me think of a line from *Angels in America*, that great play which is also about not-dying, set during the AIDS epidemic: 'KS baby, lesion number one [. . .] the wine-dark kiss of the angel of death.' I'd given up on symbols and omens by then, though I started wearing gloves after that.

My mother is still in Singapore. I send her photos of the garden and give her tours of the courtyard on WhatsApp. In the corner is the Japanese maple. She always remarks on its presence, on how well it's doing, in a voice of surprise. But as she puts it: *What's the worst that can happen? It'll grow back.*

Solas,[*] *Solace*

KERRI NÍ DOCHARTAIGH

I wish I'd known, long before now, that sowing is a way to grieve.

As hands scatter seeds into earth beneath feet, they are really sculpting loss.

With careful, repeated movements, the hands are moulding it into a thing like light on stone.

★★★

★ The Irish for 'light'.

Aibreán, April

For as long as I've been able to name the days & months – to observe their colourful, untameable lexicon; the glistening, aching cycle of seasons in this growing, dying, turning world – I have struggled with the month of April. It sneaks up on me like the name of someone long lost in the soft pink dawn of morning. It's like the shipping forecast on the radio in those moments when sleep will not come . . . I know its beauty is there, right in front of me but (for reasons too sore and echoey to give voice to) I am unable to access it. I find myself cowering silently instead; unable to hear anything above lonely gaps in between dancing words – not yet able to see the light falling on foggy place-names – the green growth that unfurled, noiselessly, inside the winter's darkling grip.

I know that for this corner of the planet April means the land's shift towards an unknowable horizon; a season, like a place, that we can never quite pin down. Spring – *well* I know – is supposed to be a time of new beginnings, of hope, of light. But the thing is, life does not quite always tick along in the way we might will it to, and many of us find ourselves – at points of the year's circle when we imagine we *should* feel this, or that – feeling another way entirely.

The most difficult things I have experienced have all happened in that fourth month, things that spoke to me in no way of rebirth or joy; things that spoke instead of loss, of endings, of grief that stung and bit. From a young age I have been the kind of person that wills away the spring; its transitional, slippery nature has always felt a bit too close to the bone for me. I'm distrustful of its liminal, hidden ways.

2020 delivered an April – to both sides of the equator – quite unlike many of us had ever known. Many of us had sensed, for a stretch of varying lengths before then, that things were shifting in the outside world. I had felt something amiss on my isolated laneway, in the very heart of Ireland, just before *Laethanta na Riabhaiche*, 'the borrowed, skinning days' that mark the end of March and the beginning of April in old Celtic lore. The news was confusing, grew harder to decipher with each passing day, and the outside world's chaos mirrored changes that had gone on beneath my own skin; in places – like the soil's strata – that you can't really see. After three and a half decades constantly on the move, from one rented space to another, a lifetime spent running with all that I could carry in my arms, I had found somewhere where I knew I'd be able to stay.

Two months before the pandemic reached us here in Ireland, my lover and I moved into a small stone railway cottage he'd been left the year before. It was a wreck, in need of more work than we were sure that we could manage – but when we'd first gone to see it, more winged creatures filled the sky above the overgrown garden than I'd ever seen before. Creepy crawlies & bats, damselflies & moths, things we didn't even have the names for. We didn't need to say out loud that we'd both fallen for the place hard. As 2019 drew to a close we sold most of our things, left all we knew behind, and hunkered down for winter on a muddy, isolated laneway in the middle of Ireland. Storms raged at the house, the garden and us for the first months. When the storms eased, the pandemic took their place in rattling the outside and inside worlds.

Lockdown came, and my lover set at the brambles and thorns in the wild space behind the house with any tool he could get his hands on, clearing and making space. Having never lived anywhere for any proper period of time, I had no idea what to do in a garden. If I'm honest, I was reticent at the start. If I'm truly honest, I was terrified. I'm someone who has had to make peace with loss; with being forced to leave space after space, places that I never really quite knew how to grieve. How does someone learn to stay? To give their body over to a small stretch of land? What does it mean to tend to a garden, when so much pain has been inflicted in our world when it comes to land?

We were not alone in throwing ourselves into the small stretch of land we could suppose to call 'our own' this year. My social media threads were full to bursting with people placing their hands into the soil, trying to quieten down the cacophony of the unfolding news. It felt as clichéd as any of us could really imagine, but as the world outside became more and more scary, rather than hiding inside, many people entered deeper into the earth's hidden folds. A funny old mix of gratitude and guilt swam inside my belly, and again I don't think I was alone. That some were stuck, lonely, surrounded only by bricks and strangers, broke my heart afresh each day. What gave me the right to stand among birds and brand-new life, green and hopeful?

I had less than no clue what to do. The when, the how, the where of it all. Despite the excitement it overwhelmed me – in ways I found too hard to really talk about back then. After weeks spent raising the handful of seeds I'd been lucky enough to get my hands on, one freak storm halfway through the first lockdown lifted my flimsy plastic 'greenhouse' and hurled it against the boiler; practically everything I'd nurtured into being was lost. I posted the pictures of the scene on Instagram, seeking some kind of response from people I mostly haven't even met. I could make no sense of it: the obsession & the torment, the desire

& the peace it all brought, somehow. The responses were all from women. Some I 'knew' already, many I didn't. Every single one of them was soothing, encouraging: full of goodness. And full of something else, too. Full of sharing. Lettuce & radish seeds arrived from Bournemouth with love and with crockery. Wildflower seeds for the bees from Brighton, deadly nightshade and an art deco postcard from France, cosmos and a hand-made ring from Bristol, honesty seeds from Orkney.

So much more than seeds was being passed between us. We shared, from our tiny, back-lit screens – in various degrees of lockdown and of heartache – tips & images, resources & timeframes; our gardens felt like almost womb-like spaces. Places we might feel held; spaces outside of the normal confines of place or time. Being there, no matter what we lost, no matter what blows we and the garden suffered, gave so many of us a means of feeling buoyed, as well as a way to feel set free, and so much less alone than we really were. We were almost on a shared allotment, or a row of gardens on the same street; something about the posting of seeds, the giving of advice, the commiseration after failure felt primal, beautiful.

I read, again and again, how people now recognised the garden as somewhere that held the potential to imagine things coming along; things they might long have taken as impossible. How it might hold you close, as you realised – through bone-achingly tiring, rhythmical work – that you felt differently about things than you had before you began.

Many of us have carried much, so much, into the hidden places within our gardens, and other spaces, in these last seasons. For an unfathomably long period of time, the only other person I set eyes upon was the one I lived with. But I felt less alone than I had for most of my life. I tried to record those early days, as best I could, as a way to remember those oddly boned days; in that year like no other.

★★★

Bealtaine, May

3rd
Sat, for the whole joy-bright day, in the garden. Papers, coffee & goldfinches for my love's birthday. No visitors to the laneway but the cuckoo and a butchering ginger tabby. Feared, again and again, for the wrens. Cried at kind message from G on Instagram, as sun set – peach streaks against faded gold. It all feels

too much, the goodness that still lives in the world, despite it all.

10th
More yellow in the garden that I have ever seen. Rereading *The Grassling*, of course.
We have decided, against every odd, to try for a baby.
Stood alone at the ash and wept. Cows across the stream, nonchalant, as ever.
Increasingly taken by them, unsure why.

Meitheamh, June

1st
Observation, the act of really looking, listening and attending to the world around us, feels necessary. Black Lives Matter movement & campaigns to protect abuse victims in media to extent never have witnessed before. The time to listen & act is now.

10th
Teilgean: cast, throw
Soilais: light
How is it that sometimes when we cast away, or throw something off, veins of illuminating light ripple through the harbour. Moved beyond measure by Nick Hand's words on Bristol tearing down that statue.

13th
Have been wakening to birds in the house, fledglings in the morning, beating against my bones: wanting to be let back out into the bright, still world.

22nd
It is the turning of the year, the still point of the summer solstice has passed; we are making our way towards the next part of the circle. Middling, gossamer time. The light out there is fading; I watch it bleed out from a new moon, solstice sky.
Particular light, though, when it comes, it comes to change us.
Some light, when it comes, it comes to stay.

Iúil, July

17th
The ash came down. In the hours just before, starlings wheeled in the sky like noisy fishwives. Cannot help but feel it all means more than it does. Planted comfrey from M's garden in ours.

19th
MD died. Don't feel in any way how I felt, for decades, that I would. Neither anger, nor relief. Died early evening of 16th.

There is so much light in the garden with the ash gone. Moths coming in such numbers as to seem almost otherworldly.

Lúnasa, August

9th
Dreamt of watching moths & nightbirds in a planet-lit pale-yellow field, from a slowly moving, lantern-soft train carriage, in a summer both later, & earlier, than this one.

10th
First day away in many months. Connemara, in the full sun of summers long passed by.
Read Woolf. Gathered pebbles for the garden.
Swam beneath murmuration of sandpipers that painted the clouds silver.
Bought test at Salthill.

11th
Poipín*
Stood just now in my first garden, surrounded by the poppies I grew from seed, with a creature inside me the size of a poppy seed.

* The Irish for poppy.

22nd

Baby now the size of a sweetpea. Ours in no hurry to stop. Can't keep up with cutting them.

Sent some home with P for his mam with a copy of Winter Papers. He was beaming.

Meán Fomhair, September

22nd, Equinox

I placed, for the first time, in the soil – seeds – & waited in a way that I,

before, had never known how.

Light – unseen, unheard, unbelievable – came from cracks; in the whole vast, quiet world

– & taught them: how to dance inside the circle of becoming.

Deireadh Fomhair, October

8th

A beautiful, rhythmical sound, like the boats, when they rattle; in the winter winds.

Heard the heartbeat for the first time.

24th
Stones & eggs,
bones & nests –
solid objects,
to be cradled in my hand,
to be weighed,
to be held up to the silent light
– to be kept

Samhain, November

13th
Collected, in the stillness of the near gloaming, what I think could be the last of the flowers. Poppy heads & exquisite, skeletal remains of love-in-a-mist, blue cornflowers fading into lilac, clover, & lavender from three different plants, sculpted quaking grasses & bright yellow fennel.

24th
Two years sober. Lay in bed watching light drip into the garden from a frosty, starry sky.
Wee creature inside flutters like a moth bird; against the window of my skin.
Should bring in the last of the kale and have with eggs.
The Blue Nile on repeat.
Howled like a baby beneath the sycamore, in gratitude.

27th
Half-way grown round. Moon above the garden, white, and mirroring.
Oak / Frost / Digging / Beaver / Mourning / Reed / Snow.
Light spills out from everywhere.

Nollaig, December
1st
The wreath is hung.
Grew, from seed, every part of it myself, in my first garden.

15th
Hydrangea, against the once-white walls; like rust, on gorgeous, abandoned objects.
The winter's beautiful bones, in the shelter of the speckled thrush.

18th
Have found myself, in the depths of the darkest half of the year, on the cusp of the turning; thinking of swallows. How they swooped and wheeled in the sky; above a garden full of wren-song. Remembering the pelvic girdle of a delicately bird-like rat, given by my lover; found as he cleared away shoots springing up beside the dead ash tree. Thinking about April's seed

pink moon, as the world ached and broke, and made room for healing.
Of how different next April might be.
23 weeks today.

20th
The longest night.
Sky all the blues that ever existed.
Moon – a startlingly beautiful crescent.
So much sorrow as everything changes yet again for folk.
Loneliness, and too much to even contemplate.
Last light before solstice.
Blush-pink rose beside lady's mantle still in bloom.
Watched Saturn & Jupiter hold each other, close as lovers.
A Christmas star, the first for centuries; above our first garden.
Wee drummer inside me, playing on my belly, as I pull out both the leeks.

★★★

I wish I'd known, long before now, that sowing is an act of trust.

That the body, as it sows & plants, when it tends land,
as it hopes for growth,
gives itself over to a vast & shifting future it could
never, in that moment, quite imagine.

I wish I'd known that to sow is to scatter light around,
like wee fluttering moths.

The Sustainable Garden

How to Weed: A Writer's Guide

ELIZABETH-JANE BURNETT

'I'll give you a choice of borders,' she instructs – *where you can't do much damage*, comes the under-breath. She takes me around the bed edges, pointing out who is for the chop: dandelion, clover and chickweed. 'I'll remove the dandelion if I can keep that moss,' I say. I feel the injustice of the promise but have a strong attachment to mosses. She consents, leading me to the top of the hill. I trudge behind, reminded of the last time I tramped after my mother in this field. I was very small and can still recall her smooth glide out through the grass, a swan leaving a slick of plants for her cygnet to inch past. As she brushes a rose bush, one of the few flowers left this winter afternoon, a few petals are dislodged. They shower behind her like confetti.

I fall to my knees in the soft wake of my mother's movement – the dropped petals, where the ground blushes pink – and make the first incision. The steel blade of the trowel pushes through the topsoil and the red burst of upturned soil makes my whole face glow. I love this earth. The Permo-Triassic rocks of red sandstone have broken down to form Devon's Redlands. As steel hits stone, I think about the movement embedded beneath: the splintering of mountain, the rise and fall of lake. Nothing about this ground is fixed, there is no such thing as a mistake.

I meet my first worm within minutes – then a slug, a snail, another worm – and place them to one side of the upended earth. My progress is slow as I stop several times to rehouse residents. I think of the last protracted period I spent along this border – the previous winter, I had charted the mosses at this spot from the top of the field down to the stream at the bottom. That walk had been a movement through grief: finding a new moss in a path well-worn, charting a new life after my father had gone. Now, there is a different loss. Work I had started had thrown me off course. After striking past topsoil I had met rock. After reaching rock, I had thought I had better stop.

When I return from fetching the wheelbarrow, I find I am no longer alone. A robin is standing in the soil. Its red breast matches the earth so that it is hard to tell where one ends and the other begins. After watching me for a while, it launches a flurry up into the air, rapidly beating its wings – small phoenix, rising from clay. The robin is often associated with the spirits of humans who have passed away. It watches me from its landing place in the laurel hedge that my father planted.

In this spot, as a child, I had watched him. In one of my earliest gardening lessons, I had followed the fork as it was driven into ground, again and again in a relentless battle that was not mine, yet, I had somehow wanted to join. With some reservations, he had handed me the fork, and, momentarily, stepped away. I had plunged the tine straight through my foot. This time, I think, I am putting the blade down, not driving it though. This time, I am not so eager for approval as to keep asking things of the body that it was not formed to do.

bell hooks, writer, teacher and author of *Teaching to Transgress*,[1] describes how, in the weeks before hearing if she has secured an academic position, she is 'haunted by dreams of running away – of disappearing – yes, even of dying'. I take a long time to re-position the trowel and shift each small body away from harm. She writes, 'These dreams were not a response to fear that I would not be granted tenure. They were a response to the reality that I would.' I curl a glove around the lower parts of the plants. It is not easy to take them apart. hooks's fear is that she 'would be trapped in the academy forever'. The taproot is hard to untether. It is like reeling a fish in. I finally move the white cord free as the robin watches from a leafless tree. The deeper the earth, the stronger the smell; peppery soil mingles with lemon balm as I fall back with the force of the uproot. I am taking more time than my mother would. 'I won't give you that border, you'd never finish,' she had said. I think of everything I would like to have left unfinished, that ever led me the wrong way or took too much: situations, people that I should have left sooner but didn't because of that need to complete which we all inherit. The word 'complete' with late fourteenth century roots, recalls the old French 'complet' – *full* – and the Latin 'completus' – *filled up.* The word, then, conveys having no deficiency, wanting no part or element. Yet how many times, for me, has it meant the opposite? Completing a task that carries no heart produces one level of difficulty. But to complete another's need is an even slower bleed to a lower horizon. It takes time to unpick desire. To lay each strand out through time and ask – *is it mine?*

Now I have found my rhythm. Pull the plant, push the worm, pull and push, push and twirl in the lengthening toil. I do not like to disturb the dandelion, the resident whose roots can sink many feet below, loosening the soil, aerating and helping to reduce erosion. The deep taproot is an enabler, bringing nutrients like calcium up from the depths and fertilising grass. Broken-off pieces of root can sprout new plants – this is the curse of a stringent weeder but a tiny comfort to me.

I do not like to cleave the clover which, like the dandelion, works to improve soil health, converting nitrogen into fertiliser using bacteria in its roots. I smooth its shoots, running through its vernacular names – Milky blobs, Sheepy-maa's, Bee-bread – and think of the bee's flights into flower for nectar. I don't like to be the threat here. Clover has several alliances – with worms, as well as pollinators. Seeds that have been buried for a long time may come to the surface in winter through worm casts. These can be used as shelter before the warmer dawn of spring. I don't want to interfere with anything.

I am loath to tear the chickweed, whose Latin name, *Stellaria media*, contains a star – 'stellar'. The small white flowers light the soil. Herbalist Juliette de Baïracli Levy noted the resemblance between uses of chickweed around the world and slippery elm, the latter a sought-after mucilage-rich digestive and emollient plant.[2] As slippery elm has become endangered in the wake of Dutch elm disease, chickweed has become more valued. Ralph Waldo Emerson said, 'A weed is a plant whose virtues have not yet been discovered.'[3] But virtue – I tussle each flower head loose – is not only tied to use. It can be virtuous, sometimes, to refuse.

'You never wanted to be a teacher,' hooks's sister reminds her. She responds, 'She was right. It was always assumed by everyone else.' I handle the strands of clover and chickweed, trying to parse out what to cut and what to leave. The stars of chickweed-light coil, like pressed flowers, bright showers of meteors boiling in earth, frothing, alert. 'Since we were little,' hooks's sister replies, 'all you ever wanted to do was write.' The words break free from the soil's stacked scree. When you listen to yourself deeply it is hard to switch off, when you pull the words out they are ashen and hot – whole planets that writhe in the dark, just waiting to be given a chance.

I reach the tiny forests of mosses. Plants that have been around for millions of years, remaining almost unchanged through each structural shift in the earth, offer, as author and academic Kimmerer has described, a different model of success: 'Not to be the biggest, the most powerful, or in control, but to have longevity in living well on the earth.'[4] hooks describes a moment where she falls, after working beyond her capacity, at the 'limits of being tired – "bone weary"'. She knows that she needs to renew but it is hard with so much left to do. I survey the sweep of the border. Piles of ejected earth, flower and leaf. Bends of moss bodies I have managed to leave. 'The academy is not paradise,' she concludes. 'But learning is a place where paradise can be created.'

The garden is not paradise. But weeding is a way of escaping – removing what you don't wish to keep, breaking the circuits you don't want to complete. There is much in the academy that the writer needs to leave. Often, in the academy, the writer is the weed. Waiting for their virtue to be seen can mean surrendering the best of their being. I greet the mosses, whose names I now know, as they gleam with green in the gladdening glow. *Thuidium tamariscinum*, *Kindbergia praelonga*. They show me their way of developing stronger. Linked to each other in earth, yet able to adapt and re-birth. I listen a while to their chattered light speech, follow the channels and eddies of their sparkling, startling bodies.

By now, my stomach muscles are starting to ache, there is a throb in my spine and I am starting to shake. Lifting up from my folded position, I dig my heels in and rise from the waist, shoulders back, head held high in space. Over the leaf of the laurel is hazel and over the hazel is radish, a leap – all vanish – in the squeak of the kestrel – the leaves of the clover join the weave of the hawk, as the dandelion cleaves its leaf into stalk – 'dandelion', from French, 'dent de lion' – *tooth of lion*, with sharpened leaves it probes the earth that it lies on. There is a pile of the torn in the barrow and a feeling of air in the marrow. I leave the job half-finished – this feels important. The barrow empties into compost, in an endless recycling of loss.

As I start to back away, a delicious sound comes from the side of my head. Tiny squeaks, chattering. It is the robin but making a noise I have seldom heard before. Neither song nor call, this is a smaller language, a little, intimate fall of notes. It is like a child, whispering in your ear, it is the most strengthening sound to hear. It hops onto the earth at my feet and launches itself into air with a flare of feather and fling. What is it doing? It feels like play. I stand back, looking at my unfinished task, unmarketable, unremarkable and start to laugh but softly, so as not to scare it, a sort of low-level chuckling sub-song. The robin continues the click and trickle of its own.

All evening I smell of soil. Despite soap and water, it is on my fingers, under my kneecaps, caught between filaments of hair. The freshness of the air just hangs there. And the petals that dropped as my mother brushed past, shimmer a moment then break in a path that has passed. Where mountains have loosened and rivers regressed, it is nothing to choose to step back, or to rest. I am full of the absence of snares. There is nothing inside me that tears. Just the sound of low song and the drift of the earth and the murmur of mosses turning into the rock and the sift of the grit in the body's small bend now it has loosened and started to mend. A small shift. A life, lifts.

Just Call Me Alan

CAROLINE CRAIG

It started as a bit of a joke. A tip shared with a friend on how to get orchids to reflower yielded miraculous results. 'Cheers, Alan,' she said. Then another friend asked me why I thought the succulents in her damp, windowless bathroom kept dying ('They're desert plants, mate . . . How about showing them the sun?'). The name, in reference to Mr A. Titchmarsh, MBE and illustrious *Gardeners' World* presenter extraordinaire, just stuck.

To give some context: I had worked at Kew Gardens promoting the research of the world's foremost plant scientists. And whilst doing a similar job for the Royal Horticultural Society, I sat in the same office as their chief horticultural advisor Guy Barter. So I can only assume that friends thought that I, through osmosis, had obtained a diploma in horticulture and metamorphosed into an Alan.

True, I have picked up a few tidbits along the way. I can tell you a little about the bee needed to pollinate vanilla orchids, what structural colour is, and I tend to waffle on about flavonoids after a few.

Somewhat hilariously, for a short time I even had an allotment in Kew Gardens, apparently entirely unfazed by competition from some of the world's best growers, a handful of whom trained alongside Alan himself. I'd finish my morning meetings and cycle the rusty team bike to a quiet, far-off corner, hidden away from the general public, where I'd spend a blissful hour tending to a tiny patch of UNESCO World Heritage Site. But today, I don't have a garden of my very own, just some pathetic windowsill plants. I'm on the waiting list for an allotment near my flat (I think), but realistically that's a while off. Am I really worthy of such a name?

In fairness, that's not the full picture. My maternal family are fruit farmers. They've been working the same fields in the South of France for generations, as subsistence farmers until the 1920s, then, embracing the changes that the mechanisation of agriculture brought about, for cash crops.

They have grown everything under the sun, from almonds to peaches, truffles to asparagus, figs to walnuts.

My entrepreneurial great-grandad Aimé even tried his hand, unsuccessfully, at rearing wild boar (clue's in the name, Papy). Today, though, it's mostly Syrah and Grenache for wine, cherries for eating and olives for oil.

I am not going to claim that this background makes me a farmer by association either, but I have spent a fair proportion of my life mucking in. I've harvested tonnes of grapes, I can prune an olive tree, make nettle fertiliser, desucker a vine . . . So while I might not be a full-blown Alan I could probably manage to hold a conversation with Him, for a few minutes at least.

Provence is another world, and our little pocket is nestled in the foothills of the Mont Ventoux. Each field, orchard or vineyard has a different name, reflecting its history, geography, shape or the name of its previous owner. La Touraine. Saquet. Gassin. Les Sept Raies. Les Quatre Chemins. It makes me laugh to think that Alexis, a cherry orchard adjacent to a stone pine forest, is still named after some bloke my great-grandad bought it off in the 1950s. There's nothing intentionally whimsical about this; it's purely for practical reasons. If you have tens of vineyards, cherry orchards or olive groves dotted around the place, you need to know which parcel of land someone is talking about. It makes conversations possible.

The parcels have been divvied up between the family over the years, but we all gather for bigger tasks and harvests when the time comes. These moments are magical. A typical day during the grape harvest, known as the *vendanges*, in early autumn, goes like this:

We meet as an extended family, early in the morning and position ourselves in the rows of vines behind a tractor pulling a trailer. We work our way up the rows, swiftly picking bunches and tipping our filled buckets into the trailer as we go.

Effective picking is about going as fast as possible, keeping up with everyone, not leaving a single bunch, and finding time to examine the grapes, clipping off any that are rotten, unripe or dried up, all of which would affect the quality of the wine.

We stop at midday to eat, either picnicking in the shade of a *cabanon*, the little stone houses that dot the Provençal landscape, or if it's too hot, we drive back to the nearest farmhouse.

On the last day of picking, we will have a celebration meal and barbecue, *des grillades*, in the vines: usually grilled merguez and lamb chops, and of course, wine, toasting another year and another harvest.

In Provence, lives move with the rhythms of the seasons, and food is at the heart of everything. A farmer's job is to feed, but that's not where it ends.

Much of life also revolves around dealing with and managing surplus fresh produce. Making coulis, compote, cordial, nougat, brining olives, even wine, are in reality exercises in prolonging the shelf life of perishable fruit.

But some surplus we create on purpose, for the simple pleasure of continuing food rituals the family has been taking part in for generations. This is where the *potager*, the kitchen garden comes to the fore. Most fruit farmers in the region will have one for household consumption; some plants are grown for cash and others for pleasure. It's just the way it is, even though it means extra work.

Walking through a fruit and vegetable farmer's *potager* is a bit like peering in a chef's fridge at home. What are they growing? What are the tricks? Is it even organic?

My uncle Tonton Serge can cultivate just about anything beautifully. But he has a head start because there is a source in his garden that gushes pure drinking water and, as if that wasn't enough, the Auzon river also runs through it. His soil is dark and fertile and he grows a beautiful array of produce for himself and for us. Our challenge is to eat it before the wild boars. My parents keep almond trees in their back garden for the sole purpose of making lavender honey nougat to eat on Christmas Day. My sister Estelle is the custodian of a special walnut tree we use to make *vin de noix* in June. My great-aunt Tata Régine was the Roma tomato queen, which in turn, made her the tomato coulis queen, as this variety is best for sauces. My mother Françoise grows Romas now that Tata Régine has passed away, and my sisters and I make the coulis for winter every summer, dutifully following our great-aunt's recipe. My grandad Papé Xime grew everything and anything but loved his aromatic herbs best and planted rows and rows of them behind his little cottage. He'd provide his sister Edmée with the bucketful of basil necessary for the annual family *soupe au pistou* gathering in high summer.

For me, this is the most evocative aspect of the Provençal landscape: not what is cultivated in vast quantities in the fields, but what is growing in the back gardens of houses. It provides a window into the culture and rituals of the region's people.

Is there anything more beautiful than a vegetable garden or an allotment? The word 'paradise' when studied etymologically means 'walled garden'. I'd take it a step further as I'm greedy, and say that paradise is surely standing in any organic fruit and veg perma-culture garden.

But this is where the subject becomes difficult for me, reconciling my love for my family, for Provence and our heritage and activity there with what I know about the untold pressures on the environment right now.

Summers in Provence are getting drier and drier, and we are all increasingly reliant on irrigation. Agriculture is both a user and polluter of precious water sources. Farmers are custodians of a large part of the planet's land surface, the people who work to feed us all. But at the same time, there is a competing environmental need to maintain the earth's biodiversity, life-giving, life-preserving and life-enhancing as it is.

We all know this. But finding a practical application in balance with economic development is a challenge. What we grow as a family in Provence is organic, sure, but that just isn't enough. If the person in the field next to you isn't farming organically, then your produce probably isn't organic either. Even if they dutifully spray crops on days when there is no wind, chemicals seep into the soil and groundwater. In fact, the soil you stand on can harbour decades' and decades' worth of residues of long-since-banned pesticides. Another issue is monoculture. Vast swathes of fields with just one cultivar planted create a strange, unnatural landscape in reality. Appearances aside, it often means there are insufficient flowering plants around to feed insects, aka the pollinators. For most edible plants, no pollination means no fruit.

Wild plants and wild landscapes must be protected at all costs. Work or spend any length of time at Kew and you learn that as well as being beautiful and important elements of ecosystems, plants are chemical factories, and that they hold secrets, cures and powers we are only just beginning to understand. Today for example, a vast proportion, almost a third, of all modern medicines are derived from plants: in essence, synthesised plant chemicals, precious beyond all measure.

Interviewed by Sam Knight for a *Guardian* long read, my former colleague and renowned botanist Carlos Magdalena explained beautifully and tragically what is lost when a plant species becomes extinct:

> Each chromosome is a letter. Each gene is a word. Each organism is a book. Each plant that is dying contains words that have only been spoken in that book. So one plant goes, one book goes, and also one language goes and perhaps a sense of words that we will never understand.[1]

The wild must be allowed to thrive alongside the cultivated, and if it is thriving, it is because we are cultivating in a sustainable way.

From the lofty heights of urban life, people, myself absolutely included, can perhaps become overly precious about all plants (or completely indifferent, of course).

Farmers often have a different attitude to things – and being the city girl that I am, it jars with me. In Provence, if a tree is in the way of a tractor turning circle it will get the chop, regardless of how many hundreds of years it has stood there. And when your livelihood depends on this year's grape crop, your organic aspirations sometimes have to be deferred if some invasive insect has moved in for a party. My sisters and I have just stopped short of staging protests at times . . .

That's not to say that my family don't care. My grandfather knew every knot of every olive tree he looked after, collected wild medicinal plants from the hills to cure his ailments, and mowed the lawn at night so as not to disturb any bees. He really, really loved nature and plants.

Nonetheless there can be a certain edge to anyone who has to actually earn a living from the land. I've seen first-hand what it can do to people and how it can twist and betray. Nature has her own ideas sometimes.

Is agriculture gardening? Technically yes, but also technically no. Is it the responsibility of the farmer to feed people or to protect the environment? It should be both, however this is a complex and nuanced issue. All life depends on plants. But a livelihood from plants today? It's not to be taken lightly.

Perhaps all of this goes some way towards explaining why I love our *potagers* and kitchen gardens so much.

They hark back to the time when we were subsistence farmers, and I see all of our future landscapes in them: a beautiful mix and mishmash of plants, nibbled by insects and animals on the edges but with enough to go around for everyone, so this doesn't matter.

People often say gardening is something you come to as you get older, that we are all Alans-in-waiting. In my case farming – or what I like to think of as custodianship – is a certainty, as I will be passed on a small share of my mum's fields in Provence and the responsibility of keeping whatever is growing on them alive. God help those plants, though they at least stand a better chance than the succulents in my friend's pitch-black bog. (I hope.)

That's a long way off though. In the meantime, I'll tend to the little pots on my kitchen windowsill in London. I called them pathetic earlier but, truthfully, I love them and they're all grown from my own experiments with cuttings. Perhaps even Alan would be impressed.

What We Know, What We Grow at the End of the World

VICTORIA ADUKWEI BULLEY

Lately, I've found myself slipping into what I have begun to call *magical apocalyptic thinking*. This is the only way to describe what feels like a distinct mode, often induced by low-hum states of dread, in which I busy myself with imagining what might be possible after the end of the world. By itself, the phrase *end of the world* is loaded. It suggests a nothingness that bears no signs of life, and so I include here with it the word *after*. I am looking for a way to speak of the world that we inhabit now, but this alone is incomplete. I want also to speak through and over it – as you might do with a wall – to what may exist on the other side. In thinking of that existence, that strange after-place, I am imagining into the spaces made available by the absence of everything that makes life barely liveable for all but a few. I want to know, with specificity, what will be needed of me for the welcoming-in of that world. I would like to be ready with the kind of skills or knowledge that will matter then.

It's hard to say what these flights of fancy achieve in real time, except that they feel urgent when they take place, and that they often relate to nature. It makes sense, then, that I have these thoughts mostly when I am outside in the garden, or walking or cycling through any local stretch of forest, or simply at the kitchen counter making a loose herbal tea. In a time during which it is necessary to ask what structures must be dismantled in order for all peoples to live freely and well, thoughts about what will need to be abolished come in tandem with those asking what we will need to learn to grow.

My mother always told me that there is a remedy for everything; that for anything that ails the body there is a herb. Like many indigenous beliefs, this concept is encoded into language. In Ga, the word for medicine is *tsofa*, which breaks down into *tso*, for tree, and *fa* for root.* In this body of thought, the challenge to healing is not the non-existence of a cure but the lack of wherewithal to locate and access it. There is always the possibility that what is needed is out there, somewhere, sprouting from the ground.

A natural consequence of growing up around ideas like this is an early sense that there is really no such thing as a weed. This has never left me: the concept of weeds has always felt somewhat fascist at core. There are no weeds, there are only spatially undesired plants that possess a capacity to survive, with no help from you, in places where you are trying to cultivate something other. Weeds are only threatening to those who insist immoveably upon the idea of control and well-kept borders, and as is the case with all language through which artificial hierarchies are expressed, the word *weed* also becomes a barrier to deeper knowledge. When a plant such as a dandelion is perceived as just a common weed, the use of its root as a liver-cleansing digestive bitter is likewise dismissed. When, as a kid, I would help my father to remove them from the garden lawn, this was the knowledge of dandelions that I didn't yet have.

* Thank you to my herbalist friend Naa Adjeley for alerting me to this.

Mostly my magical apocalyptic thinking takes the form of thought experiments. During these bouts, I pose a series of questions to myself and seek to answer them, one at a time. The questions are big and terrible to start out with, but I work them down to as best a nearness and immediacy as I can manage. I zoom in until I have a macro view of a problem and see it writ large, like a hair on the arm of a larger concern.

The questions are mundane. They run something like: *if all shops were to close indefinitely, and there were no electricity or gas, what would we do, then, in order to eat?* Like: *if we had no lighters in the house, what would we use to start a fire with only our own two hands?* They continue. To write this piece I am first using a mechanical pencil made by Pentel. It has the word *ENERGISE* printed across its barrel, and in many ways, writing does that for me; fuels me in a way that leaves me feeling sharp and deliberate. If I couldn't do it, I would want to. And so, I ask myself, with all shops shut, what I would use to write if I were out of lead and electricity, with no access to more. In such a situation what could I do, really, about anything? I picture survival TV programmes, bring to mind stones to use as flints for fire, and think of natural pigments. I picture the dried hibiscus flowers in a jar in the kitchen, brought back from a trip to Morocco. The colour released from them when soaked in hot water is rich enough to paint with. But pencils and their lead are not edible; I still don't know what I would do for food.

Before she quit the city for a farm in the countryside, our landlady – a faceless woman whom we have never met and maybe never will – lived here in the flat we rent from her. We know nothing about her other than that she must be monied and green-fingered. In the garden of this ground-floor flat are four silver birches, three apple trees, one plum tree, three grapevines – with actual grapes* – and strawberry plants in wooden boxes. At the rear of the garden is a patch of raspberry bushes, the leaves of which make a good tea. The woman also used to keep chickens.† Beyond the vegetable patch that we planted ourselves, none of this orchard-like mise en scène has anything much to do with us. And though the rent we pay to live here seems 'reasonable',‡ spaces like this are inaccessible to most Londoners.

Many of my earliest childhood memories are of being in the garden. Mixed in with these are moments I remember from before I went to school, on days when

* Mad, right?

† Again, I have no idea who she is.

‡ Nothing about the rental market in London is reasonable.

I was at home with my mother. One thing confuses me here: I did also attend nursery, so I don't know exactly when these home days were, or even if they were frequent. I only know that in these memories my older siblings were at school, and that Dad was at work, and so Mum was home with me during the day because she worked night shifts at the hospital.

Sometimes, in these memories, she is carrying me and we are looking out of the kitchen window. In one vision, we go out into the garden together with a jug of water and pour it onto a small patch of the grass. As if like clockwork, earthworms seep out of the soil, glossy and iridescent with their blue-green sheen, and I am amazed each time. It is always early morning.

In another recalling, it's autumnal outside but the lawn is alive with a flock of starlings. They work across the grass mechanically, their movements staccato and their voices a chorus of swirls and chirps that sound like old radio static. We try to count them before they fly away, but they move too fast and there are so many of them.

In my later childhood I spent time in the garden entirely of my own accord, liberally and alone, watching and digging at things. Once or twice – but never since – I found what is called a *false scorpion* under a rock and ran to show it to my mother. With her own knowledge of actual scorpions, she told me to be careful and to put it back where it came from. At other times, I tried to collect the flowers of a small plant that grew between the cracks of the garden's paving. It had conical yellow flowers that, when squeezed, released the fruity scent of pineapples. I always wanted to make a perfume from them.

When I bring to mind these memories, it occurs to me that many environmental changes are noticeable even from within the small enclosure of a garden. Starling numbers have dropped across the country, which is partly why it's such an event now to see them in murmuration.* False scorpions, I recently found out, are rare – so hard to come by that in 2016, one species was recorded in Dorset for the first time in eighty years. And *pineapple weed*, which to my grown-up joy turned out to be the actual name of the fruity-smelling herb, is not only absent from my parents' garden today, but seems difficult to come across anywhere now. It seems to have disappeared, along with its properties which are similar to those of its close relative, chamomile.

Even with all of the late summer fruit on offer here, one aspect of our garden that I continually find myself

* Starling murmurations are magical events in their own right, involving patterns of mass movement only seen elsewhere in a behaviour of atoms called *phase transition*.

most taken by is the compost bin. Maybe this is because it's the first and only one I've encountered in my life. There's a quiet excitement in me each time I lift the lid – I never know what I'm about to witness, and I'm always interrupting something. Removing the lid on any day immediately sends a whole set of invertebrates into a frenzy and I stare at it all for a moment, ignoring the smell of decay. Every visible and non-visible organism in there is busy with the work of turning something dead and rotting into something sustaining of life, and so I pour out our food waste, replace the lid, and leave them to it. I walk back to the kitchen and joke with T that it wouldn't be so bad to be reborn as an insect in a compost bin since every time I open it up it's like a rave in there.

I can almost see it, I say, *with a sign flashing K O M P O S T above the entrance in neon lights.*

So far, in our time here, we have grown a number of herbs and vegetables across seasons with varying levels of success. These include tomatoes, courgettes, peas, pak choi, beetroot, spinach, chillies, radishes, fennel, carrots, onions, basil, thyme, sage and yarrow. Some of our efforts were destroyed overnight by foxes. Of all our attempts, the courgettes and spinach have been the most rewarding.

You place something into the ground and you tend to it. You see nothing for days but you keep tending all the while and then, if the necessary conditions are right, slowly you notice something: a small pale stalk, some bright green starter leaves. Starter leaves – or *cotyledons* – are those first two little leaves you see on most plants grown from seed. You can't easily know what a plant is just by looking at these. They are not what are called *true leaves* and as such they don't bear any characteristics specific to their species. They are plant life in embryonic stage. They test the atmosphere and help the plant to know whether the world it enters will be safe enough to thrive in.

During a poetry reading in the midst of global protests following the killing of George Floyd,* I once said that black people should touch the earth as much as possible, so as to remember that it is always beneath our feet.† What I mean by this is that regardless of where we are on the earth – with our manifold and variegated histories – we have been displaced from our relation to it by forces still at work today. Whether diasporic or continentally African, whether the surviving

* Break into the Forbidden, organised by Ignota, 5 June 2020.
† I thank Daniela Valz Gen here for her work and thinking in bringing me closer to these sentiments.

descendants of enslaved peoples or surviving descendants of colonial subjects, we exist all the while *in the wake*[1] of a rupture that altered our belonging to land *anywhere*. In this sense, what I am speaking of here is not so much a question of where we are, but *how* we are – how we live now in relation to the natural environment. I am thinking with the work of Kathryn Yusoff here, who articulates how *colonial possession (of subjects, land, resources)* and *forced eviction from land* all facilitate *the disruption of ecological belonging*. All of this amounts, as Yusuff states, to *the mutilation of land, personhood, spirituality, sexuality and creativity [. . .] a process of alienation from geography, self and the possibility of relation.*[2]

And yet the land is still here. Beaten and bruised and fully remembering all that has occurred, but here nonetheless, right beneath our feet. In the wake of all that has happened in the recent history of the earth, it is easy to forget that. And so, if we can, we should hold our hands to it. Often.

A Small Herbal Inventory:
Lavender, for calm and fragrance. Lemon balm, for optimism and calm. Nettle for digestion and grounding. Yarrow for the liver, focus, emotional overwhelm and shielding. Hawthorn for the heart and heartache. Rose (nearly finished) for the heart also. Red raspberry leaf

for the womb. Hibiscus for the blood. Dried, plaited twigs of rosemary from the bush in front of the house, useful for smudging the house; cleansing, protective.

When I think of what it might mean to master something, to become an expert or elder who is learned in a way that could be useful, I often think that I would like to become a living library of the knowledge of the earth. It is true that I want, as James Baldwin once declared, to be *a good writer.*[3] But beyond that, too, I want a somatic knowledge. A knowledge that is embodied in the richest sense – one that goes beyond a Google search, or the leaves of a book, or what is inscribed on a list dictated by my grandmother who knows her herbs by use (and by sight, when she still had her vision) and not by the written word. This is a knowledge that comes from being and listening in relation; one that is earned from kinship and clan. It is a knowledge built from using one's available faculties to touch and taste – where safe – and smell. A knowledge with which one can look with the body; point with the hand and say with the mouth *here is holy basil, here is passion flower, here is goldenrod*, while there is still time.

Gardening, then, is a practice of sustained noticing. And though it should go without saying that open, natural space is something everyone should have immediate access to, gardening itself is not about having a *garden*. It is not about growing food or flowers but instead about developing our sense of how ecosystems work – and working with, not against them. And since no plant necessarily needs human help to grow if the habitat is right, it seems to me that gardening is less about growing plants than it is about growing your own understanding of how they best live. In which case, the garden is *you*.

Here is a process, a mode of connection in which we understand that growth of any kind means that you must both work and wait. Water, and wait. What happens or does not happen next will not be entirely up to you. And yet, always there is right action; right relation at stake. Always there is responsibility, or what some might call stewardship. There is the embodied understanding of one's place in the endless ensemble of living and breathing things that *love and eat one another*,[4] each in concert with the next, in infinite *one-anotherness*. Knowing all of this, I slip my hands into the darkness of the soil and all that it holds of a future. I hold the earth as it holds me, and allow myself to be changed.

About the Contributors

Victoria Adukwei Bulley is a poet, writer and artist. An alumna of the Barbican Young Poets, her work has featured widely across publications including *The Poetry Review*, the *London Review of Books* and *Chicago Review*. She was the recipient of an Eric Gregory Award in 2018, and has held residencies internationally in the US, Brazil, and the V&A Museum in London. Victoria is the director of *MOTHER TONGUES*, a poetry, film and translation project exploring the indigenous language heritages of black poets. Her debut pamphlet, *Girl B*, was released in 2017, and in 2019 she was awarded a Technē doctoral studentship at Royal Holloway, University of London, for practice-based research in creative writing.

Niellah Arboine is a writer, journalist and the lifestyle editor and a founding member at *gal-dem*. She is the author of an essay in *gal-dem*'s YA anthology *I Will Not Be Erased*, a contributor to the cookbook *Community Comfort* and has presented a BBC Four documentary on Black journalists. Her work focuses on identity and culture. She was born and raised in south London.

Elizabeth-Jane Burnett is an author, academic and founder of *Grow Your Own Creativity*. A writer of English and Kenyan heritage, she was born in Devon and her work is inspired by the landscape in which she was raised. Publications include the poetry collections *Swims* (Penned in the Margins, 2017) – highly commended in the Forward Prize and a *Sunday Times* Poetry Book of the Year – and *Of Sea* (Penned in the Margins, 2021), the monograph *A Social Biography of Contemporary Innovative Poetry Communities: The Gift, the Wager and Poethics* (Palgrave, 2017) and nature writing memoir *The Grassling* (Penguin, 2019). She leads the British Academy/Leverhulme Trust project, *Creative Writing and Climate Change: Moss, Wetlands and Women* (2018–21) and is the nature diarist for *Oh* magazine.

Caroline Craig comes from generations of fruit farmers in Provence where her great-grandfather, Aimé Rimbert, co-founded the wine cooperative in her family's native village, Mormoiron. She is the author of the Gourmand award-winning *Provence: Recipes from the French Mediterranean*, *The Kew Gardens Children's Cookbook: Plant, Cook, Eat*, and the co-author of the cookbooks *The Little Book of Lunch*, *The Cornershop Cookbook* and *The Little Book of Brunch*. She is the editor of *Metropolitan Magazine*.

Jon Day is a writer, critic and academic. He teaches English at King's College London, and his essays and reviews have appeared in the *London Review of Books*, *n+1*, the *New York Review of Books*, the *Times Literary Supplement* and many others. He is a regular book critic for the *Financial Times* and the *Guardian*, and writes about art for *Apollo*. He is the author of *Cyclogeography: Journeys of a London Bicycle Courier*; *Homing: On Pigeons, Dwellings and Why We Return*; and *Novel Sensations: Modernist Fiction and the Problem of Qualia*. His edited anthology of fishing writing, *A Twitch Upon the Thread*, was published in 2019.

Kerri ní Dochartaigh is from the north west of Ireland but now lives in the middle, in an old railway cottage with her partner and dog. She has written for the *Irish Times*, *Winter Papers*, *Caught by the River* and others. She is the author of *Thin Places*.

Jamaica Kincaid was born in St John's, Antigua. She is an award-winning writer whose books include *At the Bottom of the River*, *Annie John*, *Lucy*, *The Autobiography of My Mother*, *My Brother*, *My Favourite Plant*, and *My Garden (Book)*. From 1976–1996 she was a staff writer for *The New Yorker*. She lives with her family in Vermont.

Daisy Lafarge's first poetry collection, *Life Without Air*, was published by Granta Books and shortlisted for the T.S. Eliot Prize 2020. She has received an Eric Gregory Award and a Betty Trask Award, and was runner-up in the 2018 Edwin Morgan Poetry Award. Daisy is currently working on *Lovebug* – a book about infection and intimacy – for a practice-based PhD at the University of Glasgow. Her debut novel, *Paul*, is forthcoming from Granta Books in August 2021.

Penelope Lively is the author of many prize-winning novels and short-story collections for both adults and children. She has twice been shortlisted for the Booker Prize: once in 1977 for her first novel, *The Road to Lichfield*, and again in 1984 for *According to Mark*. She later won the 1987 Booker Prize for her highly acclaimed novel *Moon Tiger* (subsequently shortlisted for the 2018 Golden Man Booker). Her other books include *Going Back*;

Judgement Day; *Next to Nature, Art*; *Perfect Happiness*; *Passing On*; *City of the Mind*; *Cleopatra's Sister*; *Heat Wave*; *Beyond the Blue Mountains*, a collection of short stories; *Oleander, Jacaranda*, a memoir of her childhood days in Egypt; *Spiderweb*; her autobiographical work, *A House Unlocked*; *The Photograph*; *Making It Up*; *Consequences*; *Family Album*, which was shortlisted for the 2009 Costa Novel Award; *How It All Began*; and the memoirs *Ammonites and Leaping Fish: A Life in Time*, and *Life in the Garden*. She is a popular writer for children and has won both the Carnegie Medal and the Whitbread Award. She was appointed CBE in the 2001 New Year's Honours List, and DBE in 2012. Penelope Lively lives in London.

Claire Lowdon's novel *Left of the Bang* was published by Fourth Estate in 2015. She was Assistant Editor at *Areté* for eleven years. She has written for *Areté*, the *Sunday Times*, *TLS*, *Spectator*, *New Statesman*, *Literary Review*, *The White Review* and others.

Paul Mendez was born and raised in the Black Country. He now lives in London and is studying for an MA in Black British Writing at Goldsmiths, University of London. He has been a performing member of two theatre companies, and worked as a voice actor, appearing on audiobooks by Andrea Levy,

Paul Theroux and Ben Okri, most recently recording Ian Wright's *A Life in Football* for Hachette Audio. As a writer, he has contributed to the *Times Literary Supplement*, the *London Review of Books* and the *Brixton Review of Books*. His debut novel *Rainbow Milk* was named one of the *Observer*'s top 10 debuts of 2020.

Nigel Slater is the author of a collection of best-selling books and presenter of nine BBC television series. He has been a food columnist for the *Observer* for twenty-eight years. His books include *The Christmas Chronicles*, *Appetite*, the critically acclaimed two-volume *Tender*, *The Kitchen Diaries I, II* and *A Year of Good Eating: The Kitchen Diaries III* and *Eat: The Little Book of Fast Food*. His award-winning memoir *Toast: The Story of a Boy's Hunger* won six major awards and is now a BBC film starring Helena Bonham Carter and Freddie Highmore. *Toast* has also been adapted for stage by Henry Filloux-Bennett and directed by Jonnie Riordan. In 2020 Nigel was awarded an OBE for services to literature and cookery.

Zing Tsjeng is the executive editor of VICE UK and the author of the Forgotten Women book series. She specialises in women's and LGBTQ rights, politics, culture and lifestyle.

Francesca Wade has written for publications including the *London Review of Books*, *Times Literary Supplement*, the *Paris Review*, *Granta* and *New York Times*. She is the winner of the Biographers' Club Tony Lothian Prize, and a recipient of a Robert B. Silvers Grant for Work in Progress and a 2020–21 Fellowship at the Leon Levy Center for Biography. Her first book, *Square Haunting*, was longlisted for the Baillie Gifford Prize. She lives in London.

Notes

A Common Inheritance

1. George Orwell, 'As I Please', *Tribune*, 18 August 1944
2. Todd Longstaffe-Gowan, *The London Square* (Yale UP, 2012)
3. George Augustus Sala, *Gaslight and Daylight, with Some London Scenes They Shine Upon* (Chapman & Hall, 1860 [1858])
4. Octavia Hill, *Homes of the London Poor* (Macmillan, 1883)
5. Julian Wolfreys, *Dickens's London: Perception, Subjectivity and Phenomenal Urban Multiplicity* (Edinburgh University Press, 2015)

Companion Planting

1. Rebecca Solnit, 'Revolutionary Plots', *Orion Magazine*, July 2012

Five Tongues

1. Jamaica Kincaid, *My Garden (Book)* (Farrar, Straus and Giroux, 2001)

How to Weed: A Writer's Guide

1. bell hooks, *Teaching to Transgress* (Routledge, 1994)
2. Juliette de Baïracli Levy, *The Complete Herbal Handbook for Farm and Stable* (Faber and Faber, 1991)

3. Ralph Waldo Emerson, 'The Fortune of the Republic', lecture delivered at Old South Church, 30 March 1878
4. Robin Wall Kimmerer in Janice Lee, 'An Interview with Robin Wall Kimmerer', *The Believer*, November 2020 https://believermag.com/an-interview-with-robin-wall-kimmerer/

Just Call Me Alan

1. Carlos Magdalena in Sam Knight, 'Why would someone steal the world's rarest water lily?', *Guardian* long read, October 2014

What We Know, What We Grow at the End of the World

1. Christina Sharpe, *In the Wake: On Blackness and Being* (Duke University Press, 2016)
2. Kathryn Yusoff, *A Billion Black Anthropocenes or None* (University of Minnesota Press, 2018)
3. James Baldwin, *Notes of a Native Son* (3rd edn, Beacon Press, 1992)
4. Linda Hogan, *Dwellings* (W. W. Norton, 1995)

Daunt Books

Founded in 2010, Daunt Books Publishing grew out of Daunt Books, independent booksellers with shops in London and the south of England. We publish the finest writing in English and in translation, from literary fiction – novels and short stories – to narrative non-fiction, including essays and memoirs. Our modern classics list revives authors whose work has unjustly fallen out of print. These books are printed in striking editions with introductions by the best contemporary writers. In 2020 we launched Daunt Books Originals, an imprint for bold and inventive new writing.

www.dauntbookspublishing.co.uk